T0234007

Data-Driven Engineering Design

Ang Liu · Yuchen Wang · Xingzhi Wang

Data-Driven Engineering Design

 Springer

Ang Liu
School of Mechanical and Manufacturing
Engineering
UNSW Sydney
Kensington, NSW, Australia

Yuchen Wang
School of Mechanical and Manufacturing
Engineering
UNSW Sydney
Kensington, NSW, Australia

Xingzhi Wang
School of Mechanical and Manufacturing
Engineering
UNSW Sydney
Kensington, NSW, Australia

ISBN 978-3-030-88183-2   ISBN 978-3-030-88181-8   (eBook)
https://doi.org/10.1007/978-3-030-88181-8

This Springer imprint is published by the registered company Springer Nature Switzerland AG
The registered company address is: Gewerbestrasse 11, 6330 Cham, Switzerland

Contents

Chapter 1
Data-Driven Engineering Design

Abstract This chapter aims to introduce the basics of data-driven engineering design to design researchers and practitioners who are unnecessarily familiar with relevant notions. The background, motivation, and significance of data-driven engineering design are elaborated. From a theoretical perspective, data-driven engineering design is characterized by a novel integration between the existing design theory and methodology with the emerging data science. Therefore, a systematic design process is divided into multiple key design operations, and a complete data lifecycle is divided into multiple data operations. A theoretical framework of data-driven engineering design is presented to couple various design operations with relevant data operations for different scenarios of engineering design.

Keywords Engineering design · Design theory and methodology · Data science · Design process · Data-driven design

1 Introduction to Data-Driven Design

In light of the sweeping trend of digitalization and cyber-physical integration in every facet of modern society, data is becoming the new 'oil' in the digital world. Identical to the critical roles of oil in generating electricity, driving machinery, and powering vehicles in the physical world, data is projected to play equally, if not more, important roles in constructing new knowledge, commanding smart devices, and facilitating decision-making in the digital world.

To be fair, even before the big data era, data has always been an integral part of engineering design. However, the conventional practice of engineering design has been to extract relevant information from raw data, construct design knowledge based on information, employ knowledge to solve design problems, and accumulate design wisdom/insight through cognition. In the past, a number of highly influential design theories and methodologies (DTM) have been developed to facilitate the back-and-forth transformations of design data, information, knowledge, and insight [1]. Since the 2010s, largely as a result of the advent of Industry 4.0 [2], the process and practice of engineering design are becoming more digitalized than ever before.

Data science is an emerging field that investigates the approaches, algorithms, and technologies to obtain knowledge, understanding, and intelligence based on data. As a transdisciplinary subject matter, data science integrates scientific advances of mathematics, statistics, information science, computer science, and so forth. Practical applications of data science can be found in many industries such as e-commerce, manufacturing, health care, finance, transportation, energy, etc. The emergence of data science puts forward an appealing possibility for engineering design, i.e., how, in what ways, and to what extent can data be directly employed to drive design, especially in the context of new product development? On the one hand, conceptually identical to 'water', data naturally flows into every step of an engineering design process as well as every phase of a product's lifecycle, hence paying the way for data-driven design. On the other hand, conceptually similar to the challenges of utilizing marine power where energy is produced by small-scale changes, data can only benefit engineering design if a massive volume of data can be collected, transmitted, stored, cleaned, processed, analyzed, visualized, integrated, and protected in a theoretically sound, methodologically systematic, practically scalable, and contextually customizable manner.

Data-driven design can, therefore, be regarded as an innovative strategy of integrating the longstanding design theory and methodology (DTM) with the emerging data science, where the latter is leveraged to enhance the applicability, effectiveness, and adaptability of the former in the big data era. Data-driven design is applicable to many design activities in product development, such as interpreting customer voices, generating design concepts, identifying design couplings, personalizing product specifications, managing design complexities, enhancing product quality, and providing value-adding services. Data-driven design can empower manufacturers to join forces two types of 'power sources', design data and designer cognition, to jointly drive design decision-making.

The rest of this chapter is organized as follows. Section 2 reflects the historical evolvement of the traditional design theories and methodologies. Section 3 encapsulates the current development of data science with respect to the data facet as well as the scientific facet. Section 4 presents a structured data-driven design framework that is intended to support the integration between DTM and data science. Section 5 summarizes this chapter and previews other chapters.

2   Design Theory and Methodology (DTM)

2.1   Historical Reflection of DTM

Before the popularity of design theory and methodology (DTM) within the engineering domain in the 1970s, the subject of design was investigated primarily as a sub-field of art. The pioneering work of Herbert A. Simon, *the Science of the Artificial* [3, 4], paved the way for the subsequent advancement of DTM over half a century. Since then, the focus of engineering design shifted towards the synergized development of descriptive design research (i.e., describing why, how, and in what fashions a desirable, functional, and feasible artefact is designed and redesigned) and prescriptive design research (i.e., prescribing a systematic design process or pattern that can be followed by designers to design new artefacts with desirable attributes). The unique nature of engineering, which applies scientific findings to create new machinery, substance, structure, and process, makes it a perfect platform and testbed to accommodate the rigorous exploration of design.

As a result, a number of highly influential and widely adopted design theories and methodologies have been developed in the context of engineering, especially concerning mechanical engineering and production engineering. Some notable examples include, but not limited to, Quality Function Deployment (QFD), Kano Customer Model, General Design Theory (GDT), Axiomatic Design (AD) Theory, Function Behaviour Structure (FBS) ontology, Design Structure Matrix (DSM), Complexity Theory, Analytic Hierarchy Process (AHP), Analytical Target Cascading, Adaptable Design, Concurrent Engineering, Collaborative Engineering, TRIZ (Theory of Inventive Problem-Solving), Systematic Approach of Design (by Pahl and Beitz), Robust Design, Emergent Synthesis, design for X, design of X, and so forth.

Despite their natural differences, those established design theories and methodologies, more or less, share certain common scientific and methodological cornerstones. Firstly, they all endeavoured to describe or prescribe systematic processes that can be followed by designers, even novice designers, to conduct design operations in a more structured fashion. Secondly, as much as they acknowledged the critical roles of subjective designer experience, creativity, and wisdom in creating innovative design solutions, they regarded information and knowledge as the key drivers of systematic design processes as well as effective design outcomes. Next, many DTMs (e.g., Axiomatic Design, Design Structure Matrix, and Quality Function Deployment) regarded design couplings (i.e., explicit or implicit relationships between design entities in the same or different domains) as a key factor of design quality, reliability, robustness, and complexity. Lastly, they acknowledged that not only design is by nature an iterative process, but also design innovations tend to be a result of back-and-forth design iterations.

2.2 Design Operations

Different design theories and methodologies have varied terminologies to represent, classify, and synthesize design activities/operations. In this book, *design operation* is defined as an organized activity that is intended to purposefully impose desirable influences on why, how, in what ways, and to what extent an artefact (i.e., product, process, or service) is to be designed and redesigned. Accordingly, a complete engineering design process can be regarded as a series of interrelated design operations that are to be conducted for different purposes, by different stakeholders (e.g., customer, designer, service provider, etc.), with different priorities, in different fashions (e.g., independently, sequentially, concurrently, or collaboratively), and at different design stages.

In this book, the theoretical framework of Axiomatic Design Theory has been followed to represent, classify, and associate various design operations involved in an iterative engineering design process. According to the Axiomatic Design, numerous design entities can be accommodated into four independent domains, namely, customer, functional, physical, and process domains [5]. Depending on the mappings across the four domains, a typical engineering design process can be divided into three stages, namely *functional design* (i.e., the transformation from customer domain to functional domain), *conceptual design* (i.e., the transformation from functional domain to physical domain), and *technical design* (i.e., the transformation from physical domain to process domain). Each design stage accommodates a variety of different design operations.

In the *functional design* stage, the aim is to formulate a promising design target based on market intelligence, customer voice, and competition benchmarking. The inputs and outputs of the functional design stage are customer needs (CN) and functional requirements (FR), respectively. Functional design can be regarded as a purposeful 'translation' process, through which, ambiguous and inconsistent customer voices are translated into explicit and consistent functional requirements that can be accurately comprehended by engineers. A typical functional design process involves design operations such as soliciting customer voices, understanding customer preference, extrapolating customer needs, benchmarking analysis, product STP (i.e., segmentation, targeting, and positioning), functional decomposition, functional modelling, managing design constraints, specifying evaluation criteria, etc.

In the *conceptual design* stage, the aim is to develop a set of new design concepts that can satisfy the above-formulated functional requirements and comply with relevant design constraints. It is a common consensus that conceptual design represents one of the most important yet challenging design stages, as measured by its significant impacts on design cost, innovativeness, quality, robustness, etc. The objective of conceptual design is to arrive at a novel and viable design concept that can not only satisfy functional requirements and design constraints but also be manufactured and serviced effectively. The conceptual design stage comprises design operations such as concept generation, concept specification, concept modulization, concept organization, concept evaluation, concept selection, concept visualization, concept improvement, feasibility analysis, industrial design, and preliminary component design.

In the *technical design* stage, given a chosen design concept, the aim is to determine technical details and specifications (e.g., geometry, material, component, and structure) that are required to produce the concept in high quality, massive scale, and/or customized order. The outputs of technical design in turn become the inputs of production in factories. The technical design stage involves design operations such as design analysis, detailed component design, computational analysis, design optimization, production planning, rapid prototyping, virtual prototyping, simulation, design drawing, bill of materials (BoM), design for X (e.g., design for manufacturing, remanufacturing, sustainability, and recycling), design of experiment, and testing. In particular, many previous studies on computer-aided design (CAD) and computer-aided manufacturing (CAM) are intended to enhance design operations included in the technical design stage.

It should be noted that engineering design concerns the design and redesign of artefacts, which can be materialized as any 'man-made thing' that are purposeful and functional. As such, designers can follow a systematic design process to design a new product, process, or service with contextualized variations. This book focuses primarily on product design (as opposed to process design or service design), including both consumer products and industrial products in different industries, which orients the scope of this book.

2.3   Data in Design Theory and Methodology

In the context of engineering design, data can be interpreted as a form of values concerning quantitative and qualitative variables, which can characterize a to-be-designed artefact (e.g., product, service, and process) with respect to its specific requirement, constraint, functionality, behaviour, structure, health, lifecycle, etc. The simple units of data can be synthesized towards the entity of information, and a collection of related information constitutes the basis of knowledge (i.e., the perception, comprehension, and familiarity about an artefact). As such, it can be argued that the abstraction levels as well as the complexity degrees of 'data', 'information', and 'knowledge' should be organized in an ascending order.

Data is by nature an indispensable piece of engineering design. In the current practice, however, the majority of design processes, especially at their early stages, are not directly driven by data. A common practice has been to convert raw data to design information, extract design knowledge based on information, and construct design wisdom from knowledge. Although such a strategy was proven effective in the past, it is exposed to several disadvantages in the big data era. The linear conversion of 'data → information → knowledge → wisdom' is by nature complex and time-consuming, hence slowing down the cycling of design decision-making. Moreover, the iterative nature of engineering design further escalates the complexity of data conversion, which is especially true when the raw data happens to be unstructured. As data becomes exponentially bigger in terms of volume, value, velocity, and variety, it is infeasible, if not impossible, to merely depend on incremental improvements of the conventional paradigm to accelerate the conversion efficiency, productivity, and effectiveness.

Therefore, it calls for new thinking, strategies, and approaches to directly employ data to drive an engineering design process by effectively integrating design operations with data operations. As a sub-field of the Science of the Artificial, engineering design is born to be an inter-disciplinary subject matter. Design researchers have a longstanding tradition of adapting the scientific advances from other disciplines to support, streamline, and reinvent the design process and operations. A notable example is the inter-disciplinary integration between engineering design and cognitive science. Under the umbrella of design cognition, numerous efforts are devoted to investigating topics such as expert novice distinction, design decision-making, design preference aggregation, representation of design knowledge, design by analogy, design sketching, design reasoning, design fixation, and so forth. The investigation on design cognition can be regarded as the integration between design operations and cognitive operations, in which, the latter are leveraged as means to enhance the productivity, creativity, and efficiency of the former. Similarly, data-driven engineering design can be modelled as a systematic integration between design operations and data operations.

Despite the promising prospect, it should be made clear that data-driven engineering design is not an entirely new topic purely from the research point of view. Above all, one of the scientific cornerstones of data science is statistics, the key notions, principles, and approaches of which have been widely adopted in engineering design, for instance, to improve product quality, cost effectiveness, and reliability. Moreover, the emergence of data science is closely entangled with the advancement of information and communication technologies (ICTs), which are vital to the practical implementation of various data operations. As it happens, designers have an established track record of incorporating ICTs into relevant design operations (e.g., concept documentation, visualization, and analysis). ICTs can therefore bridge design operations and data operations. Last but not least, data science cannot be deployed effectively without being situated in a data-rich context with necessary enablers of domain-specific knowledge. In the past, many researchers have attempted to integrate separate design operations with relevant data operations, leading to a number of scattered studies of data-driven design.

3   Emergence of Data Science

3.1   Background of Data Science

The scientific foundation of data-driven engineering design is data science, the notion of which emerged against the sweeping trend of digitalization. Digitalization refers to the process of converting data, information, and knowledge into a digital format that can be reported, accessed, processed, visualized, and queried by digital means. Conventionally, the initiative of digitalization focused heavily on developing new approaches and technologies to identify, perceive, and measure objects in the physical world, followed by converting the acquired data values into digital formats. In recent years, triggered by the growing fusion between the physical and digital worlds (i.e., cyber-physical integration), the focus of digitalization gradually shifted towards modelling, simulating, analysing, monitoring, predicting, and regulating the behaviour, interaction, and collaboration of intelligent systems in the physical world through their digital counterparts in the digital world.

In the manufacturing industry, the initiative of Industry 4.0 is characterized by the immersive integration between advanced information technologies with manufacturing equipment, processes, systems, and services. Industry 4.0 aims to enhance the decentralization, intelligence, autonomy, and connectivity of manufacturing systems towards cyber-physical systems and cyber-physical production systems locally as well as smart factories globally. The immersive digitalization of the manufacturing sector, which constitutes the downstream of engineering design, paves the way for data-driven design in the upstream. Against such a background, more and more design operations are being digitalized accordingly. For example, paper-based surveys are replaced by online surveys, customer services are replaced by artificial intelligence, design documents (e.g., design logbook and concept map) are transformed from hand-written to digital formats, design drawings are digitalized by CAD, rapid prototyping is supported by CAM, design optimization is reinforced by powerful algorithms, and so forth.

As a consequence of digitalization, the notion of 'data science' is increasingly adopted and emphasized by both academia and industry. *Data Science* comprises a set of scientific theories, approaches, processes, and techniques to obtain new information, knowledge, understanding, and intelligence through raw data. Data science is a profoundly transdisciplinary subject matter that integrates scientific advances of mathematics, statistics, information science, and computer science. Due to its transdisciplinary nature, data science is greatly adaptable. Those common data approaches, algorithms, and technologies can be deployed to address different problems in various domains with identical patterns.

Figure 1 illustrates the historical evolvement of data science, as indicated by the public's searching trend based on Google. It is interesting to note that the interest level in 'data science' (as represented by the blue curve) has been growing steadily since 2013, when a series of technological breakthroughs collectively led to the

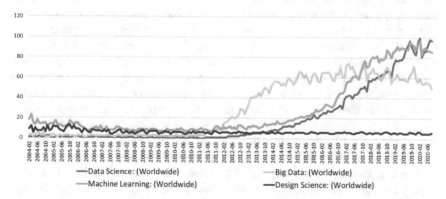

Fig. 1 Historical evolvement of data science based on Google Trend

exponential growth of data, such as the popularity of smartphones and the material-ization of Industry 4.0. It is also inspiring to notice that, prior to 2013, the interest levels in 'design science' (as represented by the red curve) and 'data science' were roughly equal. The interest level in 'data science' outperformed that in 'big data' (as represented by the yellow curve) in 2019, indicating a new consensus that it is more important to understand available data in a scientifically sound manner than simply gathering big data.

Practical applications of data science can be found in many industries. In e-commerce, online retailers rely on user data (e.g., user demographics, browsing history, searching record, customer review, customer journey, etc.) to identify target customers, recommend new products, automate transactions, and customize purchasing experience. In manufacturing, manufacturers can analyse equipment data to improve product quality, identify risks, enhance worker safety, detect anomalies, and conduct predictive maintenance. In health care, medical practitioners employ data science to interpret medical images (e.g., MRIs and X-rays), personalize drug prescriptions, and streamline hospital operations. In higher education, academics depend on student data to flip a classroom, promote adaptive learning, moderate peer learning, as well as personalize learning content and teaching pace. In finance, finan-cial institutions depend on algorithms to evaluate investment performance, assess investment risk, detect potential frauds, and facilitate transactions. In transportation, data science plays important roles in navigating self-driving cars and drones, moder-ating traffic, and regulating ride-sharing services. In the energy sector, data science is proven useful for managing smart grid, assessing energy sustainability, identifying energy saving opportunities, and predicting future energy demands.

3.2   Scientific Dimension of Data Science

The notion of *data science*, i.e., a transdisciplinary science about data, has been accepted by the scientific community merely in recent years [6–8]. The increasing volume, velocity, variety, and value of data undoubtedly contributed to the notion's popularity. However, big data by itself is not the only reason that led to the rapid emergence and wide acceptance of data science. In essence, data science is equipped with a set of interdependent abilities (i.e., descriptive, prescriptive, and predictive abilities) that comply with the commonly agreed definition of science. Firstly, data science can be used to describe and explain a phenomenon concerning a complex system (either biological system or artefact) in both physical and digital worlds. Identical to physics that can explain 'matters' and biology that can explain 'life', data science can be used to explain 'data' as well as the domain entity (e.g., a complex system or an iterative process) represented by data. Secondly, given an unanswered question or an unresolved problem, data science can be followed to inquiry the question, diagnose the problem, and prescribe new answers or solutions systematically. Lastly but most importantly, based on prior understandings of the past, data science can be followed to make reasonable predictions about the future evolvement of data as well as the domain entity represented by data [6]. The ability to accurately predict a future event is an indispensable ability for any science to be considered solid.

As illustrated in Fig. 2, data science can be leveraged to address various kinds of problems, including engineering design problems. Along the X-axis, various problems are classified into three categories, namely 'known problem', 'unknown problem', and 'emerging problem'. Along the Y axis distributes the three abilities of data science namely descriptive, prescriptive, and predictive abilities. Firstly, domain experts can rely on the descriptive ability of data science not only to address known problems and recognize familiar patterns but also to discover hidden patterns and uncover new problems. Secondly, domain experts can leverage the prescriptive ability of data science not only to optimize readily available solutions but also to propose entirely new solutions. Lastly, domain experts can rely on the predictive ability not only to predict the future evolvement of known problems but also to predict the future occurrence of emerging problems.

For decades, design researchers have devoted tremendous efforts to hopefully establishing the scientific foundation of engineering design towards the 'science of design'. A number of greatly influential design theories and methodologies (DTM) have been developed to not only describe what characterizes a good design but also prescribe how to conduct design in a systematic manner. Compared to the notion of

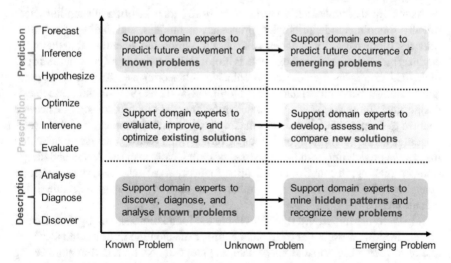

Fig. 2 Descriptive, prescriptive, and predictive abilities of data science

'data science', nevertheless, the notion of 'design science' is far less accepted by the scientific community at the moment. A longstanding bottleneck of design science lies in its lack of ability to effectively predict the future regarding different facets of design, for example, how customer preferences will change over time, how an artefact will adapt to a new environment that is full of uncertainties, what design outcomes a particular design process will yield, etc. Given the unparalleled predictive ability of data science, this book is therefore motivated to leverage the scientific foundations of data science to consolidate the foundations of engineering design.

Data has always been an important source of information in supportive of engineering design. Even before the big data era, data exists in every step of a design process, every phase of an artefact's lifecycle, and every activity of a design value chain. Despite the ubiquitous existence of data in engineering design, designers cannot always directly benefit from the knowledge that are (can be) derived scientifically from data. Before the notion of 'data science' was formally accepted by the scientific community, the path from data to knowledge was largely ad hoc (i.e., based on heuristics) or empirical (i.e., based on statistics). Data science comprises a set of scientifically sound methods, processes, principles, and techniques that can illuminate the path from data to knowledge. As such, the continuously growing design data can now be used to systematically derive new knowledge, not only knowledge about data but also knowledge about design (i.e., understandings of a design entity/process represented by data).

As previously argued, design researchers have a long tradition of adapting scientifically sound methods, principles, processes, techniques, and findings from other sciences (both natural and social sciences) to enrich the scientific foundation of engineering design. A typical example is biologically inspired design, where various biological principles are adapted to enhance engineering design through systematic analogies between artefacts and biological systems [9]. Another example is design decision-making, where the scientific foundations of decision science are adapted to enhance design decision-making [10]. In both examples, in-depth understandings of the scientific foundations of biology and decision science constituted imperative preconditions of the successes of biologically inspired design and design decision-making. In contrast, the scientific foundations of data science, as a recently emerging field, are much more unfamiliar to the majority of design researchers and practitioners.

It is therefore imperative to make visible the foundations shared by data science and engineering design. For example, a number of scientific traits of statistics (i.e., one of the scientific foundations of data science) can also be found in engineering design. Statistics played important roles in various design activities such as quality assurance and tolerance analysis. Designers rely on statistics to quantify design uncertainties, understand dependencies amongst variables in a complex system, and discover causal connections based on conditions of an effect. Better understandings of the scientific complements between data science and engineering design can enable designers to view a traditional design problem from a new data perspective, reinforce the data facet of existing design methodologies, discover unknown patterns through data analytics, make more accurate predictions, and eventually drive design decision-making based on informed predictions.

3.3 Data Dimension of Data Science

Identical to the vital roles of water in the physical world, the importance of data in the digital world cannot be overstated. On the one hand, data is the fundamental substance that empowers different kinds of digital devices, systems, and services, such as e-commerce platform, mobile device, wearable device, smart product, autonomous vehicle, smart manufacturing system, digital transaction, etc. On the other hand, various digitalized objects in the digital world will continue to generate new data through their dynamic evolvement, behaviours, and interactions.

Data in the digital world is highly analogous to water in the physical world, as they share many conceptually identical properties. Firstly, identical to the cohesive property of water, data is naturally bonded to each other. For example, the data about a customer's purchase history is naturally bonded to the data about his/her demographics, preference, and income. Secondly, similar to water that is universally solvent, data can dissolve other data. For example, the location data collected by a smart product (e.g., smartphone or smart watch) can dissolve the customer data concerning customer preferences of commute, restaurant, and entertainment. Thirdly, similar to water's high heat of vaporization, it consumes great energy and complex process to convert raw data to useful information, where the latter is notably denser than the former. Lastly, identical to the adhesive property of water, data can naturally stick to any digital system and their surrounding environment.

Data lifecycle is defined, in this book, as a series of changes that a given dataset undergoes throughout a systematic transformation from data to information, to knowledge, and eventually to more informed decisions. *Data operation* is defined as a structured activity that is intended to extract information, knowledge, intelligence (i.e., a collection of information that comprises design values), and understanding from raw data. A whole data lifecycle involves many data operations. Depending on the nature of a given dataset, its lifecycle may involve different combinations of data operations. It should be noted that *big data* refers to the data that is too large to be processed by conventional data operations. In general, big data should meet the criteria of high volume, variety, velocity, and value. Despite its undoubted importance, it should be made clear that the availability of big data is not a necessary condition of data-driven engineering design. In fact, at the early design stages, it is difficult to obtain big data about target customers, functional requirements, design constraints, design concepts, etc. Multiple common data operations are elaborated as follows such as data collection, transmission, cleansing, integration, storage, analysis, mining, visualization, security, and privacy.

Data collection refers to the process of gathering qualitative and quantitative data. In the big data era, the paradigm of data collection is reshaping from manual collection to automatic collection, from pre-defined collection to proactive collection, from occasional collection to continuous collection, and from random sampling to targeted sampling. In the context of engineering design, for example, customer voices used to be collected manually by means of focus group, individual interview, paper-based survey, and ethnographic observation, etc. Recently, it is increasingly common to collect customer voices automatically with digital means such as web crawler, camera, artificial intelligence (e.g., voice recognition and image recognition), and advanced sensors.

Data transmission refers to the process of transmitting data across devices, platforms, and user/organizations. Raw data collected from an original terminal needs to be transmitted to other terminals for the purposes of storage, computation, integration, communication, etc. In the past, data transmission used to be the bottleneck of data lifecycle. However, the recent advances of new transmission architectures, technologies, and protocols make it possible to transmit data in a significantly more reliable, faster, and even nearly real-time manner.

Data cleansing refers to the process of identifying, labelling, replacing, and removing irrelevant and incorrect data from a dataset. Data cleansing is intended to enhance data quality in terms of validity, consistency, comprehensiveness, and accuracy. A typical data cleansing process includes sub-steps such as data import, merging, standardization, normalization, verification, and export. Based on the previous analogy between data and water, data cleansing in the digital world is conceptually identical to the operation of water filtering in the physical world.

Data integration refers to the process of integrating datasets collected from different terminals, sources, and channels. Data integration serves to enhance the consistency, accuracy, and relevance of data. Data integration enables users to synchronize historical data and real-time data. In practice, a data warehouse can be built to accommodate data from heterogeneous sources for data integration. In the context of engineering design, data integration is especially applicable to the data concerning customer voice, preference, and demographics. Since a systematic design process involves the integration of various design operations, data integration is therefore an indispensable component of data-driven design.

Data storage refers to the process of storing acquired data for the purposes of documentation, organization, query, and retrieval. Traditionally, data is mostly stored locally through exclusively owned servers in a centralized fashion. In recent years, due to the increasing development of cloud computing in terms of performance, reliability, and security, more and more enterprises are motivated to relocate data storage to the cloud. As a result, the operations of data storage and data transmission become greatly coupled. The cloud-based data storage is especially beneficial for the small and medium-sized enterprises that cannot afford to invest in dedicated and expensive data storage capacities.

Data analysis refers to the process of analysing a given dataset in order to validate a new hypothesis or uncover an unknown pattern. Data analysis can be performed in different fashions with different purposes. Based on the analysis of historical dataset concerning a known event, predictive analysis functions to predict the event's future development or picture the emergence of an unknown (but identical) event.

Exploratory analysis functions to validate a hypothesis regarding a known event by summarizing the key characteristics of the event's relevant dataset. Regression data analysis functions to discover, estimate, and quantify relationships between variables. Data analysis can be used to address problems such as filtering (i.e., identifying data instances that meet certain criteria), correlation (i.e., identifying relationships between two or more variables), clustering (i.e., discovering clusters of data instances with similar attributes), anomaly detection (i.e., identifying outliners that are significantly different from others), and ranking (i.e., rank-ordering multiple data instances according to pre-defined metrics).

Data mining refers to the process of extracting hidden, implicit, unknown, and abnormal patterns in a dataset. Given some raw data, data mining plays a critical role in obtaining key information that can influence decision-making. It should be noted that there exists a subtle difference between data analysis and data mining. Given an input dataset, the former focuses on validating a hypothesis through deductive and inductive reasoning, whereas the latter focuses on discovering an unknown pattern through abductive reasoning (i.e., an intelligent guess of why). In contrast to data analysis that is commonly used in design analysis, data mining is a more unfamiliar operation to design practitioners. Traditionally, the task of extracting unknown patterns is mostly performed by designers based on human reasoning as opposed to by machines based on artificial intelligence.

Data visualization refers to the graphical representation, organization, and illustration of not only raw data but also the outcomes of data analysis and data mining. Data visualization plays an important role in facilitating human decision-makers to understand complex patterns and make informed decisions. Data visualization is long existing before the popularity of data science. It involves mappings from data values to graphic entities, structures, and representations. The underlying assumption of data visualization is that human operators tend to make more informed decisions with the support of visual cues and graphic marks, which is repeatedly proven true in cognitive science. The information graphics for data visualization include, for example, sunburst chart, heat map, parallel coordinates chart, Sankey diagram, directed network, etc. It is interesting to point out that concept visualization through CAD is intended to make visible the specifications of a design concept in terms of materials, structures, and behaviours. It can be argued that concept visualization is by nature a data visualization process, in which, the data values of design entities (e.g., functional requirements, design parameters, and design constraints) are instantiated by graphic marks.

Data security refers to the process of protecting data by preventing unauthorized user/device/agent from accessing data through prohibited means such as cyberattacks. Data is by nature a multi-faceted notion. If treated as a resource, data can be consumed to drive smart devices, information systems, and decision-making. On the other hand, if treated as an asset, data has a natural property of values for exchange and transaction. The security of high-value dataset must be effectively protected. Typical approaches of data security include, for example, data encryption (i.e., data can only be accessed and modified by users with correct encryption keys), data masking/obfuscation (i.e., some sensitive data is purposefully hidden by covering it with additional masks), data destruction, etc. Since design data tends to be associated with an artefact's intellectual properties, the values of certain design data cannot be underestimated, and the importance of data security cannot be overstated. In particular, based on the Smiling Curve Theory in business management [11], design-related activities/operations contribute great values to a new product/service, which is especially true in the high-technology industries.

Data privacy concerns whether, in what ways, and to what extent sensitive data generated through privately owned devices or personalized services can be shared with and analysed by third parties. The European Parliament and Council enforced the General Data Protection Regulation (GDPR) in 2018. Compared to the Data Protection Act in 1998, GDPR enables individual users to selectively exclude their personal data from being shared with and analysed by third parties. Certain data (e.g., medical, political, financial, and educational data) involves especially high concerns on privacy protection. For the dataset whose ownership is controversial, designers must be especially cautious in terms of whether, in what ways, and to what extent the data can be utilized in product development without compromising data privacy in an iterative design process that involves numerous stakeholders. On the one hand, various individual datasets should be aggregated and analysed, in a collective manner, to achieve the most effective outcomes of data analytics and machine learning. On the other hand, it is equally important to enable certain users to selectively exclude their data from being accessed by third parties. Such a dilemma led to new machine learning approaches such as federated learning that is designed to protect data privacy by architecture [12].

4   New Paradigm of Data-Driven Engineering Design

4.1   Definition of Data-Driven Engineering Design

Data-driven engineering design is defined, in this book, as a collaborative synergy between design operations and data operations, where the latter served as a means to arrive at a more informed, accelerated, and rationalized end of the former. As shown in Fig. 3, the integration between design operations and data operations couples a design process with a data lifecycle.

Data-driven engineering design involves tailoring data operations to the suitability of a particular engineering design scenario. It should be made clear that not every design operation nor data operation is equally applicable to a given design scenario. Practically viable applications of data-driven design should be characterized by a seamless integration between design operations and data operations. For example, the analysis of user generated contents (UGC) involves data operations such as data

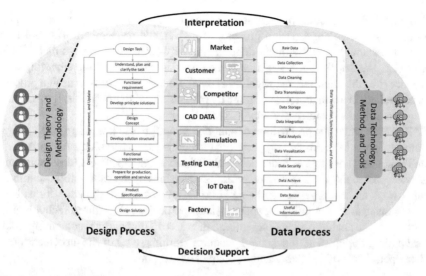

Fig. 3 Coupling between design process and data lifecycle [13]

collection through web crawlers, data analysis through machine learning, data storage through cloud computing, and data visualization.

As shown in Fig. 4, the functional modelling methodology of IDEF0 is followed to model a typical data-driven design process with respect to its input, output, mechanism, and control. The inputs of data-driven design are various kinds of datasets concerning different factors (e.g., customer, context, and constraint) that influence where, by whom, why, how, and in what ways an artefact is demanded, designed, utilized, personalized, reconfigured, and serviced. The outputs of data-driven design are desirably more informed decisions concerning different design operations (e.g.,

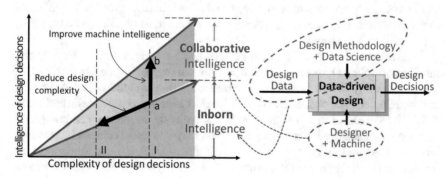

Fig. 4 IDEF0 modelling of data-driven design

problem definition, concept generation, concept evaluation, cyber-physical integration, etc.) involved in a design process. Moreover, the mechanism of data-driven design is characterized by the in-depth fusion between existing design methodologies (i.e., the scientific investigation of engineering design) and the emerging data science (i.e., the scientific investigation of data). Finally, a data-driven design process is controlled, in a collaborative fashion, by both human designer and design machine (e.g., an artificial design assistant).

The essential value proposition of data-driven design is to leverage the increasing values of data to facilitate corporate organizations (e.g., manufacturers, suppliers, and regulators) as well as individual stakeholders (e.g., customers, designers, and service providers) to make more informed decisions with respect to how and in what ways an artefact could and should be designed and redesigned. Various data operations can be supported by specific data approaches, algorithms, technologies, and tools. The enabling technologies (e.g., Internet of Things, cloud computing, edge computing, digital twin, blockchain, machine learning, etc.) of the aforementioned data operations will be elaborated in Chap. 9 of this book.

4.2   Human-Machine Collaboration in Data-Driven Design

Designers can greatly benefit from data science, in particular, its solid scientific foundation. The descriptive ability of data science makes it possible to describe, clarify, and explain design-related phenomena, such as why some customers prefer certain product features, why certain products are more complex than others, why certain structures are more prone to quality defects, etc. The prescriptive ability of data science makes it possible not only to prescribe personalized design configurations, but also to prescribe different design strategies, methods, and processes to different designers (e.g., expert designer and novice designer) for different design scenarios. Most importantly, the predictive ability of data science makes it possible to predict the future evolvement of a complex artefact with respect to its requirements, functions, behaviours, and structures.

In addition to the support for designers, data science paves the way for engaging intelligent machines to play more important roles in engineering design that has long been dictated by designers based on human cognition, creativity, and experience. Before the emergence of data science, design data has been prepared, formatted, structured, and visualized in a way that can be conveniently consumed, comprehended, and communicated by human designers. Nowadays, as data becomes increasingly large, unstructured, heterogeneous, and multidimensional, it is no longer possible for designers to independently synthesize and analyse design data solely based on human cognition.

Traditionally, engineering design is largely an experience-driven process, in which, expert designers acquire information through observations and leveraged previous experiences to solve emerging problems and create new solutions. Because of the natural complexity of experience accumulation, designers can only improve a design process, share best practices with novice designers, and customize design specifications in a highly subjective manner and at a greatly slow pace. Moreover, as design involves the creation of new ideas, methods, and artefacts, creativity has long been a research focus of engineering design. It is a common belief that creativity is one of the most effective catalysts of design innovation. Nevertheless, a longstanding challenge of creativity-driven design is to effectively reflect, duplicate, and multiply a creative design process. It is therefore challenging to generalize common design principles, to transfer creativity from creative designers to novice designers, and to map an unknown design task to known tasks.

In the traditional paradigm, a typical path of data → knowledge is to formulate a design-related hypothesis by designers based on observations, obtain relevant data through experiments or simulation, validate the hypothesis based on data analytics, and construct new knowledge based on validated facts. Design knowledge refers to the information required to understand a to-be-designed artefact. Apart from some generic design knowledge concerning the design process, most design knowledge should be specifically tailored to a particular design scenario. In the event of knowledge-driven design, the driving force of a design process remains to be human designers. In the past, many efforts have been devoted to developing expert systems that are intended to simulate how and in what ways expert designers search, retrieve, and adopt knowledge in a design process. The advances of data science make it possible to employ machines to raise new hypotheses that tend to escape a designer's cognitive attention. Different from the traditional path in which machines are commanded by designers to answer the question '*which dataset satisfies a predefined condition or known observation*', data science enables machines to potentially answer the question '*which unknown patterns characterize a given dataset*'.

Data-driven endeavours are characterized by the integration of computational thinking, analytical thinking, and systematic thinking. Compared to the traditional design thinking, data-driven design should focus on the incorporation of computational thinking into a design process. Computational thinking refers to the process of engaging computers and machines to identify design opportunities, solve design problems, and perform design operations that are once undertaken by human designers. In that sense, data-driven design differs from traditional design practices that are driven by designers, especially expert designers, primarily based on their subjective experience, knowledge, and creativity that can be hardly duplicated.

In summary, data-driven engineering design should be conducted through new forms of communication, coordination, and collaboration between human designers with intelligent design systems. Inspired by the definition of 'manufacturing system' [14, 15], a design system can be defined as a combination of human means and digital means (e.g., digital agent, device, algorithm, process, and service) that are bounded to perform a shared design task. Figure 5 illustrates different kinds of collaboration patterns between human designers and digital design systems. Depending on the specific design scenarios, an intelligent design system can serve as 'tool', 'assistant', 'peer', and even 'advisor' of human designers.

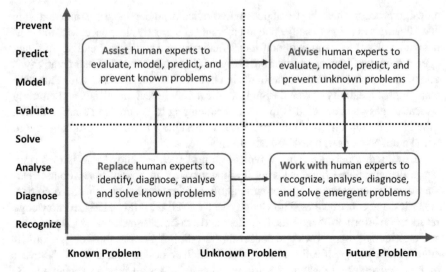

Fig. 5 Different collaborations between designers and design machine

4.3   Suitability of Data-Driven Design

Data-driven design is by no means a universal formula that is equally applicable to every scenario of engineering design and product development. As a matter of fact, data-driven design represents a double-edged sword. On the one hand, it is useful for enhancing productivity, stimulating innovations, and informing decision-making. On the other hand, it introduces additional difficulties and costs such as communication barrier between engineers and data scientists. The effectiveness of data-driven design, to a great extent, depends on bridging scientifically sound and practically feasible mappings between design operations and data operations.

The complexity of data-driven design should not be underestimated. It represents a transdisciplinary strategy that must integrate relevant knowledge, information, and methodology of multiple disciplines. Besides, data-driven design involves iterative mappings not only across different disciplines but also amongst numerous design entities (i.e., function, constraint, behaviour, and structure). Not even to mention that design practitioners and data scientists tend to follow different reasoning to conduct research and practice. Therefore, the success of data-driven design hinges not only on the integration of knowledge but also the fusion of reasoning and thinking.

Data-driven design can be oriented from either the problem space or the solution space. The problem-oriented design focuses on reformulating an existing design problem from the data perspective in terms of how and in what ways new data operations can be leveraged to formulate, interpret, and contextualize a design problem. In contrast, the solution-oriented design focuses on digitalizing a design solution (e.g., system, device, and process) to make it more capable of yielding new data that can reflect the solution's current states and predict its future evolvement. In contrast to the traditional design practice, the unique opportunities of data-driven design lean more towards informed problem definition, more immersive cyber-physical integration, and more accelerated design iterations.

Above all, data-driven design enables manufacturers to target, formulate, and reformulate design problems in a more informed fashion. Traditionally, engineering design has long been treated as a problem-solving endeavour, where designers concentrate on searching, determining, and adapting the most relevant engineering knowledge to solve given design problems. In the current practice, design problems are oftentimes defined by senior managers and chief engineers in consideration of market intelligence, previous experience, and personal insights. In sharp contrast to problem definition, problem-solving in engineering design has a long history of being facilitated by data operations such as data analysis, data mining, and data visualization. One promising research direction therefore lies in data-driven problem definition, i.e., how to explore the values of data to identify target customers, recognize anomalies and novelties in customer voices, discover unknown patterns and couplings, formulate new hypotheses about a known problem, and so forth.

Furthermore, as the sweeping trend of digitalization is transforming every facet of modern society, another opportunity of data-driven design lies in cyber-physical integration. During the past two decades, engineering design has evolved from a purely physical activity to a profoundly cyber-physical endeavour. More and more digital technologies are incorporated into the design process such as digital twin, virtual reality, Internet of Things (IoT), blockchain, etc. Traditionally, engineering design means creating tangible systems (e.g., machinery, vehicle, and device) in the physical world. The popularity of Internet, Internet of Things, mobile devices, and smart products resulted in numerous intangible artefacts such as computer software, mobile applications (APP), and web-based services. In the future, the focus of design lies in creating new cyber-physical artefacts that can facilitate human users to travel and interact in between the physical and digital worlds. Figure 6 illustrates, from the conceptual perspective, the values that can be generated from cyber-physical integration between the physical and digital worlds.

Lastly, engineering design is by nature an iterative process that involves back-and-forth iterations of design tasks, operations, and stages. Traditionally, since design decision-making has been driven by human designers based on subjective experience, knowledge, and creativity, it is difficult to accelerate design iterations (arguable any cognitive iterations) purely based on a designer's expertise. In contrast, the data-driven paradigm is characterized by a self-reinforcing mechanism: more iterations lead to more data, more data augments machine intelligence, and augmented intelligence accelerates iterations [16]. As a non-directional entity, data is by nature more suitable to accommodate iterations.

Figure 7 shows three classes of innovations that can be triggered by data-driven design. The classification is made based on the mapping between design operations (as 'ends') and data operations (as 'means'). Continuous innovation means that the same kind of data operation and its enabling technologies are employed to support the same kind of design operation. Continuous innovation involves few new mappings between data operations and design operations. For example, an algorithm can be

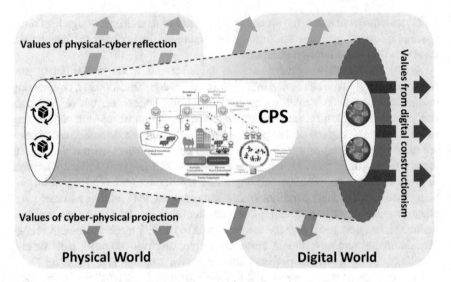

Fig. 6 Values of cyber-physical integration

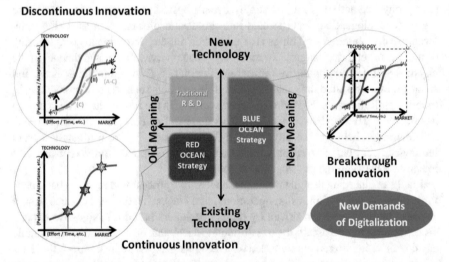

Fig. 7 Different kinds of innovations of data-driven design

continuously improved to solve an existing design optimization problem towards higher efficiency. In contrast, discontinuous innovation means using an entirely new data operation and/or its enabling technologies to support the same design operation. For example, cloud computing can be used to replace local servers for design documentation and collaborative design. Even though cloud computing and local server share highly identical functionalities, they are characterized by entirely different technological architectures (i.e., decentralized architecture vs. centralized architecture).

The most exciting innovation opportunities of data-driven design lies in breakthrough innovations, which refer to those new design operations that arise as a result of design digitalization, cyber-physical integration, and human-machine collaboration.

5   Conclusion and How to Read This Book

In summary, this chapter serves to make visible to the readers how data-driven engineering design differs from the traditional design thinking, practices, and process. A set of key notions are defined, such as design process, design operation, product lifecycle, data lifecycle, and data operation with respect to their corresponding roles in data-driven engineering design. Next, the theoretical cornerstones of data-driven design are elaborated, i.e., the existing design theory and methodology as well as the emerging data science, in regard to their historical development, practical applications, research priorities, and mutual complements. Finally, a new theoretical framework of data-driven engineering design is presented, complemented by extensive discussions on the applicability, opportunities, and innovations of data-driven engineering design.

Regarding the scope of this book, it should be noted that not all propositions are equally applicable to every design scenario. For example, Chap. 2 elaborates how to employ machine learning and artificial intelligence to analyse user generated contents and interpret customer voices. Such an approach is more applicable to the innovative design of consumer products (as opposed to the redesign of industrial products), for which, abundant user generated contents are readily available. Chapter 8 elaborates how to employ digital twin to support design decision-making through real-time simulation, monitoring, and control of a physical system throughout its whole lifecycle. Such a DT-driven approach is more applicable to the design scenario that involves highvalue assets (i.e., aircraft engine, smart city, and energy infrastructure).

The rest of this book is organized as follows. Chapter 2 prescribes a framework that can support designers to understand customer voices by analysing a variety of user generated contents. Chapter 3 explains how the conventional process of conceptual design can be made more data driven. Chapter 4 elaborates the important roles of data in supporting designers to manage constraint, complexity, and contradiction involved in an engineering design process. Chapter 5 presents a smart customization framework that leverages the values of data to empower design customization, personalization, and adaptability. Chapter 6 presents a systematic process of data-driven design of smart products with context-smartness, network-smartness, and service-smartness. Chapter 7 explains how to exploit data to integrate product and service towards more intelligent product-service systems. Chapter 8 presents a digital twin driven design framework, through which, a design process can be made more cyber-physical integrated. Chapter 9 presents a collection of specific enabling technologies that can be integrated towards a system of systems for data-driven engineering design.

References

1. Tomiyama, T., Gu, P., Jin, Y., Lutters, D., Kind, C., & Kimura, F. (2009). Design methodologies: Industrial and educational applications. *CIRP Annals, 58*(2), 543–565.
2. Lasi, H., Fettke, P., Kemper, H. G., Feld, T., & Hoffmann, M. (2014). Industry 4.0. *Business & Information Systems Engineering, 6*(4), 239–242.
3. Simon, H. A. (1988). The science of design: Creating the artificial. *Design Issues*, 67–82.
4. Simon, H. A. (2019). *The sciences of the artificial*. MIT Press.
5. Suh, N. P. (2001). *Axiomatic design: Advances and applications*. USA: Oxford University Press.
6. Dhar, V. (2013). Data science and prediction. *Communications of the ACM, 56*(12), 64–73.
7. Blei, D. M., & Smyth, P. (2017). Science and data science. *Proceedings of the National Academy of Sciences, 114*(33), 8689–8692.
8. Provost, F., & Fawcett, T. (2013). Data science and its relationship to big data and data-driven decision making. *Big Data, 1*(1), 51–59.
9. Shu, L. H., Ueda, K., Chiu, I., & Cheong, H. (2011). Biologically inspired design. *CIRP Annals, 60*(2), 673–693.
10. Scott, M. J., & Antonsson, E. K. (1999). Arrow's theorem and engineering design decision making. *Research in Engineering Design, 11*(4), 218–228.
11. Shin, N., Kraemer, K. L., & Dedrick, J. (2012). Value capture in the global electronics industry: Empirical evidence for the "smiling curve" concept. *Industry and Innovation, 19*(2), 89–107.
12. Yang, Q., Liu, Y., Chen, T., & Tong, Y. (2019). Federated machine learning: Concept and applications. *ACM Transactions on Intelligent Systems and Technology (TIST), 10*(2), 1–19.
13. Wang, X., Wang, Y., Tao, F., & Liu, A. (2021). New paradigm of data-driven smart customisation through digital twin. *Journal of Manufacturing Systems, 58*, 270–280.
14. Suh, N. P., Cochran, D. S., & Lima, P. C. (1998). Manufacturing system design. *CIRP Annals, 47*(2), 627–639.
15. Fujimoto, T. (1999). *The evolution of a manufacturing system at Toyota*. Oxford University Press.
16. Tao, F., Qi, Q., Liu, A., & Kusiak, A. (2018). Data-driven smart manufacturing. *Journal of Manufacturing Systems, 48*, 157–169.

Chapter 2
User-Generated Content Analysis for Customer Needs Elicitation

Abstract With the rapid growth of information technology (IT), user-generated content (UGC) has become a critical data resource for assisting customers in making purchase decisions. UGC also contains invaluable information for product designers on customers' behaviours, needs and opinions. In this chapter, a UGC analysis framework is proposed for researchers and practitioners to extract large quantities of information from UGC data related to product features, usage contexts, and customer opinions. This information enables designers to quantitatively measure and predict customer satisfaction levels for specific product features. The framework also provides insights into product usage contexts, which facilitates discovering product improvements and innovation opportunities. A case study of an e-reader demonstrates the application of the proposed framework.

Keywords Product design · User-generated content · Artificial intelligence

1 Introduction

With the rapid growth of IT and e-commerce, UGC has become an essential information source to assist customers' online purchase decisions. Customers tend to refer to other customers' online reviews to evaluate whether the products and services meet their needs. More importantly, many customers believe UGC to be as trustworthy as recommendations made by acquaintances. Therefore, the high-value customers place on UGC demonstrates its potential to understand user preferences, monitor customer behaviour, and gauge customer satisfaction, especially for small and medium-sized enterprises (SMEs) whose sale channels are primarily reliant on e-commerce platforms [1]. According to axiomatic design theory, the design process starts with eliciting CNs [2]. Conventionally, this process involves manual methods, such as customer surveys and interviews, to collect customer data in a structured and quantitative manner. However, UGC is unstructured, qualitative, and open-topic data, which cannot be processed using conventional methods on a large scale.

Against such background, this chapter outlines a set of guidelines for the computational analysis of UGC. By integrating artificial intelligence (AI), machine learning

(ML) and data mining technologies, the proposed framework can covert large quantity of UGC data into insightful information. The remainder of this chapter is organized as follow. Firstly, a comprehensive definition of UGC and its data sources are provided. Secondly, we introduced natural language processing (NLP) and discussed the computational model behind NLP. Thirdly, analytical models such as entity and sentiment analysis, and a social network analysis (SNA)-based importance rating approach are detailed. These models represent the foundational framework for UGC analysis, systemising the collection, pre-processing, analysis and visualization of UGC data for design purposes. Finally, a case study of an e-reader illustrates the framework's application to real-world design. Through the computational analysis of UGC, designers obtain revolutionary new ways of identifying CNs and improving product designs.

2 User-Generated Content (UGC)

2.1 Introduction to UGC

UGC is defined as media content posted by web users online, such as customer reviews, product blogs and social networking interactions. Due to the high level of creativity and descriptive detail contained in UGC, designers are increasingly using this data as an alternative to customer surveys. Product-related UGC contains valuable information about users' experiences with products, different usage contexts, preferences, expectations, and opinions on a product's quality. The ability to synthezise this data offers designers new methods for CNs elicitation, evaluating product competitiveness and market trends.

The primary advantage of UGC is that it significantly reduces the time and effort required to obtain CNs. Conventionally, conducting market research is a relatively time-consuming process, as developing surveys and sample plans at scale require costly incentives to encourage customer participation. In contrast, a large quantity of UGC is available online for free, containing up to date and objective data on customers' preferences, expectations, and opinions. Many products listed on e-commerce platforms have tens of thousands of reviews written by direct users from all over the world. As market competition becomes increasingly fierce, designers must adapt to constantly evolving CNs and preferences. Hence, UGC-driven design advantages are becoming more relevant for the design process.

Design researchers must address several challenges for the effective utilization of UGC in product design. The first issue is that UGC is an unstructured, heterogeneous, qualitative, and open-topic data form. Such data format makes it difficult to quantify and extract the information necessary for design decisions. The second issue is that UGC varies in quality. For example, one customer may not provide any valuable insights or use correct language, whilst another may offer a large amount of detail that is not relevant for design decisions. The third issue with UGC is that it ignores customer demographics who do not use e-commerce websites, which could result in a sampling bias. For example, previous research indicates that aged people are less willing to explore e-commerce websites [3], leading to the missing of UGC of elderlies.

2.2   UGC Sources for Designers

Three primary sources of UGC apply to product design: customer reviews (CR), professional testing and discussion forum data. Each source is defined, and the differences between them are discussed below.

Customer Reviews (CR) are reviews that customers of products and services post on e-commerce platforms. It is common for online shopping websites to invite verified purchasers to write about their product experiences. These reviews help potential customers evaluate whether a product will meet their expectations. Compared with other UGC sources, the main advantages of CRs are that there is a significant data quantity available and the speed at which it is updated. For example, on the large e-commerce platform Amazon.com, a product may have more than 10,000 CRs, with new data generated daily. However, not all customers provide valuable or relevant insights into a product's experience, especially in contrast to professional testing.

Professional testing data is generated by knowledgeable third parties who post product testing results on social media platforms such as YouTube. Many experienced product testers often review products on behalf of manufacturers. These reviews often contain a high degree of detail about product functions and performance and direct comparisons with competing products. However, an issue with professional testing is that they typically occur only at a new product's release date and do not consistently update with the latest information.

Discussion forums consist of dialogues between product users who are passionate about a product or hobby. In hobby forums, designers often find helpful information from customers to resolve technical product issues. The high degree of knowledge and detail provides designers with new product improvement strategies. Furthermore, members of discussion forums often have shared interests and characteristics, which helps designers understand the specific needs of particular target audiences.

3   Enabling Technologies for UGC Analysis

In this section enabling technologies for UGC analysis are discussed in detail to provide researchers and practitioners with technical references for future applications. As shown in Fig. 1, these technologies include natural language processing (NLP), computer vision (CV), and social network analysis (SNA). As mentioned, UGC data includes customer reviews, professional testing, and discussion forums, which occurs in various formats, such as text, numerical value, image and video.

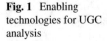

Fig. 1 Enabling
technologies for UGC
analysis

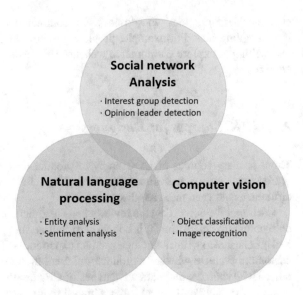

Due to the high variability between these data forms and their level of structured-
ness, artificial intelligence (AI) technologies, such as NLP and CV, are vital tools
for processing this highly complex data. Besides, as many UGC are posted for the
purpose to promote friendships and social connections, SNA is adopted to help
designers analyse such social network relationships. For example, a specific group
of customers' common interests and opinions.

3.1   Natural Language Processing (NLP)

NLP uses computational methods to understand, learn and produce human language
[4]. It can derive grammatical structure and meaning from text-based data automatically through ML algorithms. Due to advances in NLP and ML, computers can
automatically extract insights from substantial data sources, and effectively self-learn
as they process new incoming data. More NLP-based applications have emerged in
recent years, such as speech recognition, language translation and language generation. In design research, entity analysis and sentiment analysis are two critical NLP
applications for extracting valuable information from large volumes of UGC.

Entity analysis involves identifying and categorizing product features, such as
their functions or service, and product usage contexts, such as their location, weather,
or user demographic characteristics. Grammatical rules contained in UGC text identify user sentiments towards product features and usage contexts. Ireland and Liu
indicate that most of the entities that occur in UGC text are nouns or noun phrases;
therefore, enabling entities to be identified [5]. Suryadi and Kim proposed three
syntactic rules to enable designers to identify product usage context from UGC: (1)

Sentences contain the word 'usage'. (2) Sentences that contain words 'use', 'used', 'uses', or 'using'. (3) Sentences that contain the word 'for' whilst not contain the words queried by the former two rules [6]. Often multiple synonyms are used to describe the same entity; for example, display/screen, delivery/shipment. Therefore, the NLP must identify synonyms and categorize them into relevant classes. Entity analysis assists designers by distilling large amounts of complex unstructured UGC data into specific, quantitative insights. Consequently, designers can identify important insights relating to frequently discussed product features, the user's intention for purchasing the product and the environment in which products are used.

Several algorithms enhance the efficiency and effectiveness of entity analysis. The term frequency-inverse document frequency (TF-IDF) algorithm is useful for evaluating an entity's uniqueness relative to words in other documents. The TD-IDF algorithm consists of two separate equations: TF and IDF [5]. TF provides a weighting to entities that frequently occur in a UGC document. IDF counterbalances the weighting of terms that are not specific to that document by comparing the document of interest with a database of other UGC documents. As a result, frequently occurring but unimportant entities such as 'product', 'service', and 'experience' are filtered from the analysis, identifying the most relevant central topic entities. The TF-IDF equation is shown below:

$$\text{TFIDF} = \text{TF} \times \text{IDF}$$

$$\text{TF} = \frac{f_a}{f_r}$$

$$\text{IDF} = log_e(\frac{N}{N_a})$$

Here f_a represents the entity's frequency, with a being the word that appears in a UGC document. f_r is the total number of identified entities from the UGC document. N is the total number of documents, and N_a is the number of documents that contain the entity a. TF-IDF filters unimportant entities from large UGC documents.

Also, the cosine similarity measure (CSM) measures the similarity between two UGC documents. It achieves this by converting words into vectors. As all the vectors in one document are summed, it is possible to measure similarity by comparing the cosine angles of one vector sum with another. Given two documents a and b, the similarity is calculated using the equation:

$$\text{dis(a, b)} = \cos(W_a, W_b) = \frac{\sum x_i y_i}{\sqrt{\sum x_i^2 \sum y_i^2}}$$

Here, $W_a = (x_1, x_2, ..., x_3)$ is the vector of the document a, and $W_b = (y_1, y_2, ..., y_3)$ is the vector of the document b. CSM is often used for spam filtration or detecting duplicate documents.

Table 1 Example of sentiment analysis

Sentences		Sentiment score	Sentiment magnitude
S_1	The charger cable provided does not fit on any equipment	−0.8	0.8
S_2	It is in good size so I can easily handle it, also in lightweight even with cover	0.9	0.9

Sentiment analysis, also known as opinion mining, quantifies the unstructured text's emotional polarity by weighting sentiment-bearing words such as 'amazing', 'wonderful', 'poor', or 'bad' [5]. Sentiment analysis can be divided into three different levels based on the analysis granularity: entity-level sentiment, sentence-level sentiment and document-level sentiment. Sentiment analysis enables designers to understand customer satisfaction levels of products from UGC. The output provides designers with customers' satisfaction levels for the overall product and specific product features. Table 1 illustrates an example of sentiment analysis.

Determining if a particular word or phrase relating to a product is positive or negative is a critical component of sentiment analysis. The Naïve Bayes Classifier is the most frequently used algorithm for this purpose [7]. The algorithm can be trained to calculate the probability a word is positive or negative based on how the word or phrase has been used in previous documents. The classification task is performed by the equation:

$$\mathrm{P}(document|\mathrm{v_j}) = \prod_{i=1}^{lengthoftext} P(a_i = w_k|v_j).$$

Here $P(a_i = w_k|v_j)$ calculates the probability that the word w_k belongs to class v_j. After calculating the sentiment probability of each word w_k, the overall document's sentiment is determined by the equation:

$$v_{NB} = argmax P(v_j) \times \prod P(a_i = w_k|v_j).$$

The Naïve Bayes Classifier's main advantage is its ability to become well-trained with a relatively small dataset. Apart from Naïve Bayes, other classification algorithms such as support vector machines (SVM) and neural networks are also useful for sentiment analysis.

3.2   Computer Vision (CV)

Computer vision (CV) is another AI field that applies computational methods to extract insights from images or videos. With the increasing popularity of smartphones

(a)   (b)

Fig. 2 Example of **a** image classification and **b** object detection

and the ease at which users take photos, there exists a large pool of data in the form of images in addition to text. Pictures or videos made by product users contain invaluable information about their opinions and needs. Despite the many types of CV applications, the two selected for this paper are image classification and object detection. Understanding what objects are contained with images provides designers with helpful insights into how, when and where customers use their products.

The difference between image classification and object detection is highlighted in Fig. 2. Image classification analyses an entire image as a whole and classifies it into a specific category. Object detection detects specific visual objects in digital images, which is performed by object localisation and object classification tasks. Object localisation determines the location of objects in an image, and object classification assigns this object to a unique category. Deep neural networks (DNNs) have shown outstanding performance on image classification and object detection, and are widely used by academia and industry [8, 9]. Commercial computer vision APIs, such as Google cloud vision and TensorFlow object detection API, enable designers to develop their own image classification and object detection applications, by labelling and annotating the images with bounding boxes and train machine models to perform classification functions.

3.3   Social Network Analysis (SNA)

The term social network describes how people relate to each other through interactions and shared characteristics. SNA involves the study of social networks to understand their structure and implications. In online social platforms, such as Facebook, Twitter, and Instagram, UGC often generates when people discuss a particular topic (e.g., a specific product or function). The discussion usually includes how customers engage with the market, what customers like and what customers dislike.

By depicting the topic as a network of nodes related to their generators' interests, SNA can reveal, sometimes hidden, relational patterns present within topics.

In most basic frameworks, the social network C is represented as a graph $C = \{F, E, Q(E)\}$. It has three key components: a cluster of nodes F, a cluster of relationships E, and relationship strength $Q(E)$. Given a social network that has i number of entities, then the cluster F can be represented as

$$F = \{(F_i)|i = 1, 2, 3, ..., n\}.$$

If node F_i and node F_j have a certain relationship, then $e_{ij} = 1$, otherwise $e_{ij} = 0$. Therefore, the cluster of relationships E is represented by

$$E = \{(e_{ij})|e_{ij} = 1|i, j = 1, 2, 3, ..., n\}.$$

The relationship strength of the social network is represented by

$$Q(E) = \{q(e_{ij})|i = 1, 2, 3, ..., n\}.$$

Determining the relationships between customers is a challenge. Customer relationships in online social networks could be explicit or implicit. Explicit relationships occur when customers declare the relationship publicly via the social media platform. For example, one person may 'follow' or be 'followed by' another person. Implicit relationships are not formally declared on social media platforms and are typically inferred through analysing a customer's behaviour on a platform. These behaviours include sharing posts or photos, discussing topics on other posts, tagging others in posts or who or what they are commenting on in a post. Analysis of customers' explicit and implicit relationships on social media platforms reveals their interests, clusters of customer demographics, buying behaviour or personality type. Two forms of SNA are of high interest to designers: community detection and opinion leader detection.

Community detection groups customers with similar characteristics and identifies the total number of these groups within the social network platform. This macroscopic view of customer demography forms the basis of market segmentation. UGC from social media platforms contains considerable information on customers' interests. By extracting key interest terms and words from UGC, it is possible to identify unique interest categories and construct profiles of individual customers. Community detection categorizes individuals into groups based on their shared interests and observes distinctions between them.

Opinion leader detection involves finding individuals on social networks who exhibit a high degree of influence over other users. An opinion leader is a person on a social network who display the degree of influence over other users. Influence on social media equates to having a high number of connections to others who regularly view or comment on their posts. Due to their reach and status, many of their ideas and perspectives actively influence their followers. Opinion leaders who express opinions on certain product, or who have influence over specific demographics, provide designers with invaluable market insights for product improvement. Identifying opinion leaders is achieved via a metric called degree centrality, which correlates the number of connections a user has with their degree of influence.

Degree centrality measures the position of a node is within the overall structure of the social network. The degree centrality of a node is the number of relationships that the node has. The central nodes refer to the nodes that actively interact and relate with other nodes. In contrast, intermediate nodes represent the entities that lie at network boundaries, which are less related. The degree centrality can be represented by

$$D_i = \sum q(e_{ij}) \times e_{ij}.$$

D_i represents the degree centrality of node i. However, as D_i can be affected by data volume, to avoid bias, the equation is standardized:

$$D_i^{'} = \frac{D_i}{max\left(D_j | j = 1, 2, 3, ..., n\right)}.$$

Here, the degree centrality of the node i is divided by the maximum degree centrality of all nodes. Through this approach, the $D_i^{'}$ will be scaled to the range of (0, 1).

4   UGC Analysis Framework

This section presents a framework for integrating NLP, CV, and SNA techniques for processing UGC data and producing insights. The framework aims to understand a product's usage context, quantify the level of customer satisfaction it provides, determine the relative importance of its features and evaluate market trends and structures. The framework contains six key modules, as shown in Fig. 3. These include modules for data collection and preparation, text-based feature and context extraction, satisfaction analysis, image-based context detection, importance rating and data fusion. Each module's technical features, functionality and application processes are explained in detail below.

4.1   Data Collection and Preparation Module

This module's purpose is real-time data collection and increased data processing efficiency of UGC. The module operates in two stages, data collection and data preprocessing. Data collection functions to identify and collect UGC data on web pages. It uses a web crawler to detect tabular or listing type data on the page and save a

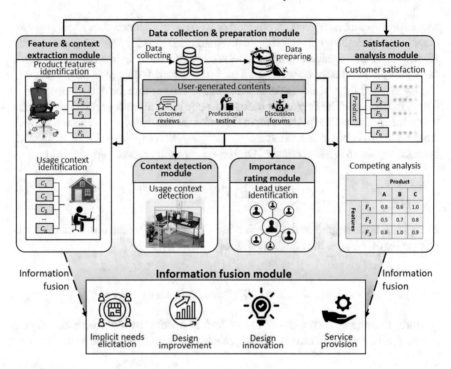

Fig. 3 UGC analysis framework

copy of data into files. Data preparation functions to clean, and pre-process data. In this stage, the irrelevant and duplicate UGC are firstly removed. The irrelevant or duplicate data can be identified using TF-IDF and CSM algorithms. Next, several NLP tools are adopted to pre-process UGC. Firstly, UGC is tokenised into a set of sentences. Next, stop-words (e.g., 'a', 'the', 'is') and non-alphanumeric symbols (e.g., '?', '&', '@') contained in the sentences are removed. Stop-words are the most common words that do not add much meaning to the sentence, which can be safely removed to improve computing efficiency. Then, numbers and letters contained in the sentence are transformed into a standardized format. For example, numbers are transformed into number signs, plural nouns are transformed into singular nouns, and letters are transformed into lower case. Finally, frequently occurred word pairs are joined with the character'_'. For example, in consumer electronics, 'touch screen' is treated as a combined word pair, 'touch_screen'.

4.2   Feature and Context Extraction Module

The feature and context extraction module identifies and summarizes the product features and usage contexts mentioned by customers in text-based UGC. The module

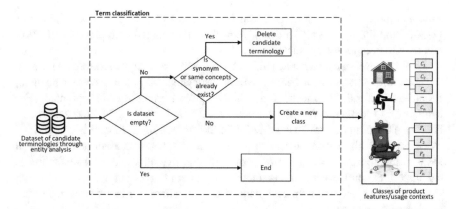

Fig. 4 Terminology categorisation flowchart

extracts nouns and noun phrases from UGC, as typically, these refer to product features or the environments in which they are used. This module outputs a summarized list of product features and usage contexts through a set of classification processes.

Considering there exists a diverse range of terms or synonyms for the same product feature or context; for example, the terms 'screen light' and 'screen brightness' are different terms that apply to the brightness of a smartphone screen, another functionality of this module is to categorize different terminologies that used to describe the same feature or context. Figure 4 below illustrates a process for categorizing similar terminologies. A sample of identified terminologies forms the training dataset of candidate terminologies. And each candidate is assigned into classes by answering two questions: (1) Is the dataset empty? (2). Are synonyms or the same concepts of the candidate terminology already exist in one of the classes? A new terminology class is created if the candidate or its synonyms are not in any existing classes. Otherwise, the candidate feature is deleted from the dataset. The process is repeated till the dataset is left empty. This approach enables designers to build a comprehensive list of product features or usage contexts that are mentioned in UGC. It can also integrate with supervised ML to realize the automatic classification of terminologies in a large scale. The synonyms can be encoded either through WordNet software or manual labelling, and classification is achieved through Naïve Bayes Classifier.

4.3 Satisfaction Analysis Module

The satisfaction analysis module measures the level of customer satisfaction of specific product features mentioned in UGC. The emotional polarity of these features, determined by weighting sentiment-bearing words associated with them by how positive or negative they are, is an effective method for quantifying customer satisfaction.

This module consists of two steps: the first measures the customer satisfaction of features and the second involves analysing how these satisfaction levels of feature compare to competitor products.

First, the level of customer satisfaction with features is calculated by categorizing the semantics of sentiment-bearing words associated with them as either positive, negative or neutral. Such words may include 'good', 'great', or 'disappointed'. For example, in the sentence 'I am not happy with this screen light', the sentiment-bearing phrase 'not happy' associated with the 'screen light' feature is categorized as negative. If all sentiment-bearing words are assigned to and summed for each feature, the semantic strength of a word and the overall customer satisfaction is calculated. Second, customer satisfaction of the study product against competing products is measured, to help designers determine where its market advantages and areas for improvement exist.

4.4   Context Detection Module

The context detection module aims to determine the surrounding environments in which products operate by analysing digital images and videos. Enabled by CV technologies, the module can estimate the possible usage contexts, such as whether products are used indoors, outdoors, in an office, in transportation, or at home. The module contains two separate models for image analysis: object detection and image classification.

Object detection detects recognizable contextual objects, such as whether an object is a human, animal, piece of furniture or a consumer good. Identifying objects allows designers to understand when, where, and how customers interact with products [9]. By knowing what objects are contained in a product's operating environment, a designer can evaluate the impact of surrounding environments on product functions, such as whether different flooring materials will affect customer satisfaction about sweeping function.

Image classification aims to associate detected objects with the contexts where products are used. How often objects occur in unique environments allows designers to infer products' usage contexts. For example, some objects may indicate whether a product is used in a particular space (indoor, outdoor) or room types (bedroom, bathroom, kitchen, living room). Image classification aims to comprehend an entire image as a whole based on detected objects. Understanding where products are used gives designers new insights into how specific environments influence customer preferences.

4.5 Importance Rating Module

This module rates the relative importance of product features by measuring customers' preferences about product features through SNA. The importance rating is crucial for resource allocation for product design. Improving a few essential product features usually can boost overall customer satisfaction significantly. By adopting SNA, this module uses a two-step approach to determine the preferences amongst customer groups. Firstly, the individual interest network of each customer is constructed. The individual's interest network is represented by $C_i = \{F_i, E_i\}$. The cluster of nodes F_i refers to the preference tags associated with the customer, which usually are frequently mentioned product features. The relationship E_i refers to the co-occurrence of preference tags associated with one customer. Secondly, the social network is built by merging individual interest networks. The social network $C = \{F, E, Q(E)\}$ is formed by a cluster of preference tags F associated with various customers, a cluster of co-occurrences E where two preference tags are both mentioned by a customer, and co-occurrence strength $Q(E)$ that measures the number of individual interest networks that contain the E. After constructing the social network of customer interests, the relative importance of customers' preferences can be measured by computing the degree of centrality of each preference tag F.

4.6 Information Fusion Module

The information fusion module integrates UGC from various sources to produce more insightful suggestions. The multiple sources of UGC online each have their advantages and limitations. For example, discussion forum data contains detailed opinions with sparsely distributed topics. Professional testing contains highly skilled feedback but is conducted at a specific point in time and cannot be updated. Customer reviews are frequently updated, but their containing information is ambiguous and unprofessional. Integrating multiple UGC sources minimizes each source's shortcomings and enriches the quality of data designers have access to.

Despite the potential advantages of this module, there is no unified method for fusing multiple data sources. Designers should develop an appropriate data fusion model based on data availability and research aim. For example, a designer could fuse UGC from different data platforms tied to the same user. This technique provides a far more comprehensive overview of a customer's preferences than relying on a single source. Also, different data sources contain different forms of data, which each hold unique insights. For example, social network platforms may have more photos and videos of products being used. In contrast, product reviews on e-commerce websites contain descriptive texts and quantitative ratings of some product criteria. Fusing information from these data sources gives designers an enhanced overview of a product's usage context.

4.7   UGC Analysis Application

Integrating UGC into existing design methodologies can enable several promising applications for implicit CNs elicitation, design improvement, design innovation and service provisions.

Firstly, UGC analysis enables designers to elicit implicit CNs. Traditionally, often structured questionnaires were needed to obtain CNs, which limited CNs only to areas specified explicitly by designers. The major drawback of traditional survey-based methods is that they cannot identify implicit CNs, i.e., unmet or unknown wants and needs. Implicit CNs are hard to define, as these needs often change depending on the context in which products are used. Given the characteristics contained in UGCs, it is possible to extract implicit CNs. It enables designers to understand the changes of customer satisfaction in different usage contexts, such as using products for distinct purposes in various environments. Therefore, a promising application is to elicit implicit CNs in different contexts.

UGC analysis can also assist design improvements by setting benchmark performance targets. This application aims to measure the performance of a product feature against that of competing products. Such performance may include customer satisfaction, cost, and so on. UGC often contains direct comparisons between different products. For instance, in professional testing, the tester usually defines evaluation metrics and compares product performance accordingly. Analysing this data helps designers understand a product's relative performance and guide functional improvements through benchmarking.

Design innovation requires designers to develop breakthrough product functional improvements. This requirement places high demands on the designers' capability to perceive unmet CNs and select the most exciting features. To achieve this, designers must proactively explore opportunities for innovation. Such exploration includes analysing the social, economic and technological factors, for example, new technologies, market trends and social hotspots. UGC covers a range of topics that reveal new insights to customers. One example involves the detection of abnormal customer sentiment in customer reviews. The ideas or opinions raised in these reviews may facilitate discovering novel excitement functions that significantly boost customer satisfaction.

UGC analysis offers manufacturers the ability to provide more value-add services to customers. Today, many discussion forums are operated by manufacturers to support customers solve product-related issues raised in the use phase. By extracting and summarizing common issues raised in UGC, designers can develop related component warranty and replacement service. Besides, UGC realizes a more effective market segmentation, which offers great potential for designers to provide personalized service to customers. For example, designers can provide personalized product adaptation service by tracking the changes of customers' usage context.

5 Case Study

In this section, a case study of a digital reading device, known as an 'e-reader', illustrates the proposed framework's application to real-world design. Being the dominant player in the e-reader market, such as Amazon's 'Kindle', is a highly profitable and sought-after status. The reason for this is not that the e-reader itself is a profitable item; it is the repeated sale of digital books ('e-books') that produce the most revenue and profit. Analysis of e-reader UGC data offers designers new potential for improving product functions and their services.

The first step of the analysis process involved using a web crawler for collecting 5,000 CRs from online e-commerce platforms. Next, the data is cleaned and pre-processed. Then entity analysis is applied for extracting product features, functions and relevant services from the database of CRs. The training dataset was manually labelled to contain 13 feature classes and 6 service classes, as shown in Table 2.

Sentiment analysis then measures the level of customer satisfaction of specific product features and services. The sentiment polarity is scaled to a value $[-1.0, 1.0]$, with < 0 indicating a negative sentiment and > 0 a positive sentiment. The results of the sentiment analysis shown in Fig. 5 below show that customer satisfaction and

Table 2 Identified product features and service

ID	Feature/service	Explanation
F_1	Battery	Running time of e-reader before recharge
F_2	Size	Dimensions of e-reader
F_3	Apps	Applications installed on the e-reader
F_4	Price	Money paid to obtain e-reader and e-book
F_5	Charger	Charging speed
F_6	Button	An object pressed to turn on/off e-reader
F_7	Backlight	Illumination of the screen
F_8	Display	Colour, resolution, refreshment rate of the screen
F_9	Storage	Capacity of e-reader to store e-books
F_{10}	Weight	Heaviness of e-reader
F_{11}	Touch screen	The response rate of screen
F_{12}	Manual	The operational instruction of e-reader
S_1	Cover	Personalized front and back protection material
S_2	Delivery	Shipment of the product from warehouse to customer location
S_3	Prime	Prime service that consists of several services such as free two-day shipping
S_4	Recommendation	Customized advertisement made to e-reader based on users' purchase history
S_5	Replacement	Replacing dysfunctional product with a new one
S_6	Warranty	One-year limited warranty

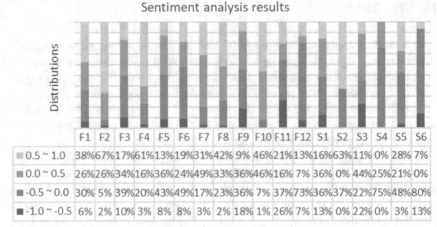

Sentiment analysis results

	F1	F2	F3	F4	F5	F6	F7	F8	F9	F10	F11	F12	S1	S2	S3	S4	S5	S6
■ 0.5 ~ 1.0	38%	67%	17%	61%	13%	19%	31%	42%	9%	46%	21%	13%	16%	63%	11%	0%	28%	7%
■ 0.0 ~ 0.5	26%	26%	34%	16%	36%	24%	49%	33%	36%	46%	16%	7%	36%	0%	44%	25%	21%	0%
■ -0.5 ~ 0.0	30%	5%	39%	20%	43%	49%	17%	23%	36%	7%	37%	73%	36%	37%	22%	75%	48%	80%
■ -1.0 ~ -0.5	6%	2%	10%	3%	8%	8%	3%	2%	18%	1%	26%	7%	13%	0%	22%	0%	3%	13%

Product features and services

Fig. 5 Sentiment analysis results

dissatisfaction opinions of features or services vary significantly. Product features such as the button, data storage capacity, touch screen and product user manual comprise > 50% negative sentiments. In contrast, other product features and services received predominantly positive feedback, such as its LED backlight, portable size and light weight.

Following the sentiment analysis, the dataset's context detection process was applied. In the case study, the product usage context refers to the product's purpose and the surrounding environment in which it is used. Extracting context-related keywords from UGC revealed the primary usage contexts to be 'gift', 'daily reading', 'trip reading'. Notably, the time of day—either 'day' or 'night'—were frequently mentioned by customers. Also, different motivations for purchasing products impact the importance of some features over others. For example, when an e-reader is purchased as a gift intended for someone else, the price has a more substantial effect on buying behaviour than whether product features will satisfy the user. Conversely, when an e-reader is purchased for reading during commuting or travel, the size, useability and battery capacity are of far greater importance than price. There exist infinite examples where product usage contexts influence customers' opinions of features. Hence, it is critical to understand customer sentiments to product features and investigate the relevant contexts that these apply.

The relative importance of product features based on their frequency is shown in Table 3. The screen backlight was the most commented-on feature from customers, which 80% of customers had positive sentiments towards. The second and third most important features were size and battery, respectively. 90% of users' sentiments were positive regarding its size, whilst only 60% of users were positive about its battery life. Price was not mentioned often by customers, yet received a very high 90% positive score, which suggests most customers feel the product is excellent value for money.

Table 3 Relative importance rating

ID	Name	Centrality	Importance
F_7	Backlight	719	High
F_2	Size	178	High
F_1	Battery	118	High
F_{10}	Weight	64	Medium
F_4	Price	37	Medium
F_8	Display	18	Low

6 Conclusion and Future Work

Increasing quantities of UGC online are enabling designers to understand customer voices with increased efficiency. Advances in AI and data analysis technologies have paved the way for a new paradigm of data-driven smart design. This chapter proposed a framework for how UGC can support design decisions. By applying the framework, designers gleaned new insights regarding product usage contexts, quantitatively measured sentiments towards specific product features, and evaluated these products' performance against their competitors.

UGC-based smart design offers three main benefits for designers. Firstly, UGC is a more cost-effective and time-saving process for understanding customer voices than traditional survey data collection methods. Secondly, a variety of AI and ML techniques obviate the manual analyses methods of designers. And finally, the large quantity of UGC available and the extraction of quantitative insights enable designers to make more objective design decisions rather than rely solely on their subjective knowledge and experiences.

Despite the successful application of the framework, two notable limitations must be addressed. First, the framework only deals with text-based and image-based UGC. However, on the websites, UGC also includes modalities such as video, and positioning data. Analysing those data are not being considered in the proposed framework so far. Second, although the framework provides a general overview of CNs, it is not precise enough to deal with the CNs of individual customers. Therefore, the UGC analysis framework is still in its early stage, but far greater potential exists if these two challenges are improved upon in future research.

Based on the research and findings of this chapter, critical areas for future research include the application of more sophisticated and robust data analysis technologies to UGC data forms. For example, the fusion technology of UGC with other data sources (sensor data, design documents, etc.) could be incorporated into the proposed framework in the future. Sensor data and design documents are effective data sources to help designers determining functional requirements (FRs) and design parameters (DPs) corresponding to CNs. In this case, finding the non-linear relationship between UGC and sensor data will be an effective way to set up the connections between customer opinions and FRs and DPs.

References

1. Lorenzo-Romero, C., Constantinides, E., & Brünink, L. A. (2014). Co-creation: Customer integration in social media based product and service development. *Procedia-Social and Behavioral Sciences, 148*, 383–396.
2. Suh, N. P., & Suh, P. N. (1990). *The principles of design* (No. 6). Oxford University Press on Demand.
3. Meng, L., Nguyen, Q. H., Tian, X., Shen, Z., Chng, E. S., Guan, F. Y., Miao, C., & Leung, C. (2017). Towards age-friendly e-commerce through crowd-improved speech recognition, multimodal search, and personalised speech feedback. In *Proceedings of the 2nd International Conference on Crowd Science and Engineering* (pp. 127–135).
4. Hirschberg, J., & Manning, C. D. (2015). Advances in natural language processing. *Science, 349*(6245), 261–266.
5. Ireland, R., & Liu, A. (2018). Application of data analytics for product design: Sentiment analysis of online product reviews. *CIRP Journal of Manufacturing Science and Technology, 23*, 128–144.
6. Suryadi, D., & Kim, H. M. (2019). A data-driven approach to product usage context identification from online customer reviews. *Journal of Mechanical Design, 141*(12).
7. Medhat, W., Hassan, A., & Korashy, H. (2014). Sentiment analysis algorithms and applications: A survey. *Ain Shams Engineering Journal, 5*(4), 1093–1113.
8. Chen, D., Zhang, D., & Liu, A. (2019). Intelligent Kano classification of product features based on customer reviews. *CIRP Annals, 68*(1), 149–152.
9. Lu, S., & Liu, A. (2016). Innovative design thinking for breakthrough product development. *Procedia CIRP, 53*, 50–55.

Chapter 3
Data-Driven Conceptual Design

Abstract Conventional conceptual design is empirical process that depends on designers' knowledge, cognition, and experience. Data make conceptual design more objective and analytical. This chapter introduces the data-driven conceptual design. There are three data services: (1) helping novice designers accumulate experience, knowledge, and cognitions. (2) Helping designers plan, redesign, and improve the design process. (3) Helping designers verify design concepts. The data-driven functional design, concept generation, concept evaluation, and affordance-based design are clarified in detail. A case study of the conceptual design for a robot vacuum cleaner illustrates the theory.

Keywords Conceptual design · Data-driven engineering · Information technology

1 Introduction

Conceptual design plays an essential role in product development by converting customer needs into design objectives and configurations, bridging customer research and embodiment design [1]. In conceptual design, designers determine the quality, cost, and fundamental features of products [2]. It is a highly innovative process in which designers employ divergent problem-solving techniques critical for the product development cycle.

Conceptual design has evolved considerably during the past three decades. In the 1980s, conceptual design followed a customer-led model whereby customers defined design objectives. Designers would then follow these objectives, formulating design configurations for customer satisfaction. During this time, conceptual design was an unadulterated empirical process, where designers relied on their knowledge, experience, and cognition for design decisions. Design outcomes were verified by physical tests and market feedback [3]. In later decades, following the development of computer-aided design and computer-aided manufacturing, conceptual design shifted towards digitalization. These two digital methods provided designers with new approaches for more efficient formulation, drafting, management, and verification of design concepts. By digitalizing design factors (dimensions, attributes, interaction mechanisms, etc.), conceptual design became more interactive and collaborative. Today, the global economy is transforming from a mode of overproduction, where product manufacturers need to achieve economies of scale, to one of on-demand production where products are tailor-made for specific markets or even individuals. The specificity in which products are made requires designers to anticipate specific customer needs during conceptual design.

Data is becoming increasingly crucial in conceptual design. Traditionally, in engineering design, data was used for design calculation, modelling, and verification. As computational power was very limited, the collection and use of large volumes of data posed a significant challenge. Developments in computer technologies spawned computer-aided design and data engineering methodologies. Here, design factors were converted into digital forms, streamlining data collection and processing. The smart manufacturing phenomenon involves the comprehensive monetarization of product data for enhancing decision-making capabilities [4].

This chapter introduces a data-driven framework for conceptual design. Subsequent sections are organized as follows: Sect. 3.2 briefly introduces conceptual design and its challenges. Section 3.3 will clarify how the services of data promote conceptual design. Section 4.4 will illustrate theories with a case study of the robot vacuum cleaner. Section 4.5 will conclude the data-driven conceptual design process.

2   Overview of Conceptual Design

2.1   Factors of Conceptual Design

Designers formulate design configurations of products based on customers' demands and expectations in conceptual design. There are eight critical factors of conceptual design, including customer needs (CNs), functional requirements (FRs), design parameters (DPs), integrated concepts (ICs), design constraints, design complexities, design contractions, and affordances. Amongst those design factors, CNs, FRs, DPs, ICs, and affordances apply to concept formulation, whilst design constraints, complexities, and contradictions apply to concept improvement. This chapter focuses on concept formulation, whilst concept improvement is introduced in Chap. 4.

CNs represent customers' expectations towards design features, services, and performances of products. They act as both fundamental principles and evaluation standards of following design stages. FRs consist of functions assigned to products to satisfy CNs. FRs are commonly in the form of 'verb + noun', and specify a target value. For example, if an oven can heat foods to a range between 100 and 230 °C, the core FR will be 'heat foods to 100–230 °C'. The generation of sub-concepts for each FR, known as DPs, provides quantitative methods for how products achieve FRs. DPs also constitute significant design factors determining a product's cost and quality [5].

ICs integrate DPs into the design configuration of products, representing their fundamental capability with reasonable technical details. Affordances are supplementary
to FRs and represent unintended uses for products.

2.2 Process of Conceptual Design

In conceptual design, designers outline product functions and generate design configurations. As shown in Fig. 1, the general process of conceptual design contains
four procedures: functional design, concept generation, concept evaluation, and
affordance-based design.

In terms of functional design, it involves formulating a list of functions to satisfy
CNs. Customer research is used for the CN generation. As textual descriptions of CNs
are sometimes ambiguous and implicit, FRs explain design objectives in a precise
and formal engineering language. FRs are classified into multiple categories, with
design strategies developed from these classifications. Engineers must ensure that
FRs align correctly with customer research. Examples of crucial functional design

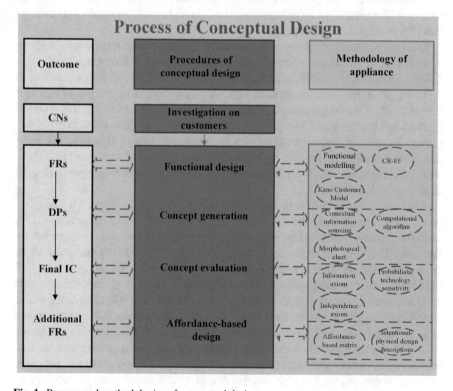

Fig. 1 Process and methodologies of conceptual design

methodologies include functional modelling, which describes flows and conversions of substances in design concept based on IDEF0 (a functional modelling methodology for systematic analysis) [6]; Customer review-driven functional formulation (CR-FF), which generates candidate FRs based on reviews of peer products [7]; Kano customer model and long-tail model, which classify different FRs based on customers' satisfaction, performance, rate of existence and popularity of usage [8, 9].

In concept generation, designers map FRs to multiple DPs. Designers' knowledge and experience are fundamental sources of DPs. Designers also investigate peer products and the voice of customers to formulate more DPs. Those DPs then integrate into the whole product's candidate concepts, which reflect essential design features. Arguably, according to Independence Axiom, each FR is achieved by a single, corresponding DP, which minimizes design redundancies, inconvenience of usage, and uncertainties [10]. However, in some conditions, engineers intend to develop multiple DPs for each FR, which is particularly important for novelty and optimal conceptual design solutions. Concept generation decisions also affect a product's overall cost, quality, and market positioning strategy. In terms of conceptual design methodologies, the Morphological Chart is a paradigm for concept and DP organizations [11]. Also, data has recently attracted more attention in conceptual design, as researchers use computational methods to generate DPs from datasets [12]. Besides, the summarization of features into different contextual information facilitates sourcing concepts and DPs via cloud computing [13].

Concept evaluation focuses on functionality and reliability of products. In terms of functionality, designers evaluate concepts via physical experiments, virtual simulations, case analysis, and their own experience. In terms of reliability, according to Independent Axiom Theory, product reliability is measured by couplings amongst FRs and DPs. Coupling formation requires strong designer intuition and is assessed by customer feedback and peer product investigations [10]. There are many methodologies for concept evaluation. For example, Axiomatic Design Theory is an important concept evaluation method by measuring uncertainties of concepts [10]. Probabilistic technology sensitivity is another critical concept evaluation approach used in mechanical design [11].

Affordance-based design is supplementary to functional design. Unlike other procedures of conceptual design, affordance-based design is led by customers. As mentioned above, affordances are ways customers' use products unintended by designers [14]. For example, the mobile phone was not initially designed to be used as a torch. However, many mobile phone users often used the screen as a light torch due to its convenience. Hence, lighting became an affordance, and now most smartphones include 'torch' as a standard function. Thus, affordance-based design is where designers explore and apply affordances to products. New approaches to affordance-based design have accelerated the discovery of affordances in recent years. For example, Jonathan and Georges proposed an affordance-based matrix that organizes affordances features and selects candidates [14]. The theory of intentional-physical design description also investigates the affordance ontology of ecological psychology through the product lifecycle. The investigation results are managed into a representational hierarchy that integrates functions and affordances into the design process [15].

2.3 Challenges of Conventional Conceptual Design

Conventional conceptual design is highly dependent on designers' own knowledge and experience. It is an empirical process where designers' knowledge, experience, and cognition determine design decisions. Conceptual design became particularly challenging for novice designers, as they lacked the critical design knowledge of experienced designers. Two significant effects occur when design depends heavily on designer knowledge. First, the productivity of conceptual design is constrained to fewer designers. Second, despite their knowledge and experience, professional designers have their own biases that impede the generation of the most novel, innovative concepts.

The evaluation of design outcomes also poses a unique challenge. Theoretically, conceptual design starts by exploring CNs, formulating FRs to satisfy CNs, and finally generating DPs to achieve FRs. CNs act as the assessment standards for conceptual designs. However, as these design factors are often qualitative and ambiguous, even experienced designers encounter difficulty determining if FRs satisfy CNs or if DPs achieve FRs. Conceptual design evaluation is a non-binary process. Concepts may achieve FRs, yet be unattractive to customers compared to competitors' products. For example, capacities of batteries are often popular topics when customers discuss smartphones. Despise always expecting a larger capacity, customers may never know the exact value they need, neither do designers. Besides, when comparing battery capacities between 4500 and 5000 mAh, everyone would instinctively prefer the larger one, whilst there might be negligible differences in practical usages. Similarly, each procedure's assessment of design outcomes poses a significant challenge.

Design concept improvement is an abstruse process. Feedback from conceptual designs may lack the precision necessary for interpreting complex engineering language and design objectives. For example, if customers complain that a product is simply too heavy, determining the precise weight of what is too heavy for a broad range of users requires a large sample. Additionally, setting a reasonable weight reduction target without compromising other feature performances creates another complex problem.

3   Data-Driven Conceptual Design

3.1   Services of Data in Conceptual Design

The importance of data is prevalent across many aspects of modern life. Different data categories such as ordinal data, discrete data, and continuous data have various practical applications. Whether in business, medical care, manufacturing, or social management, it is commonly agreed that data analysis' primary motivation is for improved decision-making. Though most conceptual design factors are tradition-ally qualitative, such data is still helpful for designers' decision-making. As shown in Fig. 2, concerning the challenges of conceptual design, there are three primary services of data:

Firstly, data help novice engineers accumulate experience, knowledge, and cogni-tions. Novice engineers often feel challenging to make design decisions due to lack of experience, knowledge, and cognitions. Engineers usually accumulate those through systematic learning, industrial work and peer communications. However, acquiring

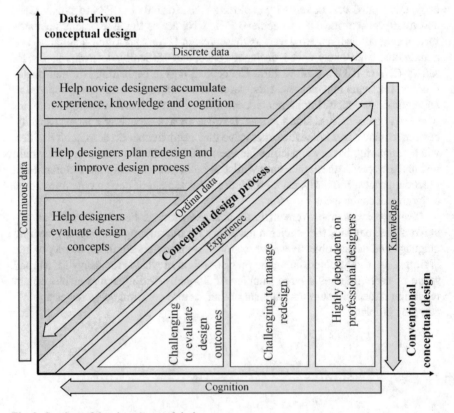

Fig. 2 Services of data in conceptual design

these skills takes substantial practice before a novice engineer becomes competent enough for a sophisticated design project. Modern data collection and analytics techniques provide a shortcut for this process. For example, data collection from product tests, real-world performance, customer feedback and behaviour, and the production process provide a wealth of helpful information for product improvement. The application of data analytics methods enables this raw data, unhelpful on its own, into useful qualitative and quantitative information [16]. This information enables novice engineers to understand their customers' needs and evaluate product performance without requiring years of experience. Thus, data enables engineers to accumulate experience, knowledge, and cognition via systematic discipline, resulting in shorter lead times [17].

Secondly, data significantly help designers with the design concept verification process. Verifications are essential to ensure consistency between functions and concepts and evaluate candidate concepts. Conventionally, design verification is a subjective process affected by designers' feelings and biases. The information obtained from data analysis transfers the features and performance of concepts into numerical language, providing valuable, objective evidence for the verification process. Furthermore, such information highlights the advantages and disadvantages of design concepts in quantitative forms. This enables designers to assess each concept's competencies against overall functionality and design goals.

Thirdly, data offer novel and revolutionary ways for design fault diagnosis and design process enhancement. Based on verification outcomes and customer feedback, strategies are developed for improving concept performance and increasing customer satisfaction. However, data obtained from design concept performance and customers are sometimes obscure and unhelpful. Therefore, it is critical to collect and process as much numerical data as possible, as the extracted information is more specific and unambiguous. Therefore, it is more common for design faults to be collected and presented in numerical forms. Through data analysis, designers obtain a deeper comprehension of customer expectations and product functionality, transpiring more objective and valuable design decisions.

3.2 Important Data for Conceptual Design and Their Collections

Conceptual design involves a wide range of different data sources and forms. Designers focus on data from customers, concepts, verifications, and peer products. Data of customers contain those that indicate customers' backgrounds, preferences, and behaviours. Examples include customers' demography, brand searches, downloads, and social media shares. Other forms of data used in conceptual design include design parameters, product dimensions, and material costs, and verification data providing feedback from physical tests or market performance. Additionally, peer product data such as shared function capability, target customer group, and physical attributes are helpful in this process. Here, peer products are defined as (1) Have similar functions or geometric design. (2) Expected to interact with the product to design. (3) Often searched together with the product to design [18]. The data collected from each of these sources is presented in probabilities, graphical distributions, frequencies, and formulae.

Data obtained during conceptual design has a variety of uses. For example, customer demographic characteristics and feedback define customer needs, scheme design objectives, and facilitate evaluation standards. Data from product testing or modelling is also helpful for estimating the product capabilities and attributes. Furthermore, concept data are also sources of DPs, as it is used in the verification process to find design faults and evaluate the reliability of design concepts. Data collected from peer products also provides an innovative source of FRs and DPs, helping designers better understand a design's pros and cons.

In conceptual design, data collection occurs from multiple sources and approaches. Standard methods for collecting customer data include direct interviews and survey. These generally extract information relating to customers' demographic backgrounds, demands, and attitudes. Designers often use online and offline interviews and surveys to maximize data collections. In an era of e-commerce, online customer reviews of products are a popular way to obtain customer feedback [19]. Existing designs and suppliers of product parts also provide valuable evaluations of current designs. Physical tests and virtual simulations are often used before they are released to the market. Generally, data from physical tests are more accurate, whilst virtual simulations are more efficient. After releasing to market, further monitoring of customers feedback provides further validation data of the design concept. In terms of data of peer products, designers can extract those data from suppliers and online. Meanwhile, designers can extract those data through the interaction of peer products (i.e., use a mobile phone to control surveillance cameras and extract data of surveillance cameras from the mobile phone). Table 1 shows some examples of data collections.

Emerging information technologies will promote the convenience and efficiency of data collection. For example, cloud sourcing and social monitoring tools help designers explore customers' demands, attitudes, and preferences. Web crawler extracts data of couplings and popularities of a design factor. Big data analytics improves quality of data and generates new data about customers and design concepts. Sensors network dynamically collect data of products and working environments in physical world. Internet of things enables products to interact and exchange data with peer products. Digital twin achieves the dynamic and comprehensive data collection from both physical and virtual world of products. Apart from those emerging information technologies, conventional approaches of data collection (face-to-face interview, manual recording, online survey, etc.) are still important in certain conditions.

As shown in Fig. 3, the following sections will introduce how data drive each procedure of conceptual design, including data-driven functional design, data-driven concept generation, data-driven concept evaluation, and data-driven affordance-based design.

Table 1 Examples of data collection with different data and sources

Types of data	Examples of data	Sources of data
Customer data	Customers' demography (education, culture background, occupations, etc.)	Customers' social media, search history on online shopping platform, online survey, etc.
	Customers' preferences (characteristics of products, brands, manufacturing variables, etc.)	
	Customers' purchase history (search history, downloads history, social shares, etc.)	
Concept data	Capability of DPs	Suppliers' resources, existing design cases, existing test outcomes, etc.
	Popularity of DPs	
	Dimensions of geometric design	
Verification data	Outcomes of physical tests	Prototype tests, computer-aided simulations, work monitoring, etc.
	Outcomes of virtual simulations	
	Market feedbacks (customers' ratings, sales amount, compliant amount, etc.)	
Peer product data	Data of functions (popularity, usage rate, reliability of functions, etc.)	Data exchange, online shopping platforms, customers' feedbacks, etc.
	Data of DPs (failure rates, usage frequencies, attributes, etc.)	
	Market performance (price, market share, rating, etc.)	

3.3 Data-Driven Functional Design

A significant amount of research explores how data-driven approaches are critical for CNs. As aforementioned, conceptual design's main motivation is developing product configurations that satisfy CNs. However, CNs vary significantly due to the different demographic characteristics of customers. Customer feedback, such as an online review or in-person interview, frequently provides ambiguous feedback regarding their products' opinions. Conventionally, designers interpret CNs through the empirical analysis of customer feedback. This methodology has three important problems: (1) Empirical interpretations of CNs are subjective, as there are significant differences between what designers think and what customers need; (2) Factors that affect CNs are miscellaneous, as designers may overlook important factors during CN formulation; (3) CNs fluctuate and change over time, and designers may not notice these changes.

On the other hand, data-driven interpretations of CNs are far more objective, as customer voices are transformed into numerical indicators. Data transfer customers' voices into numerical indicators (popularities of topics, ratings of products, frequencies of complaints, etc.) that help designers formulate, classify and interpret CNs.

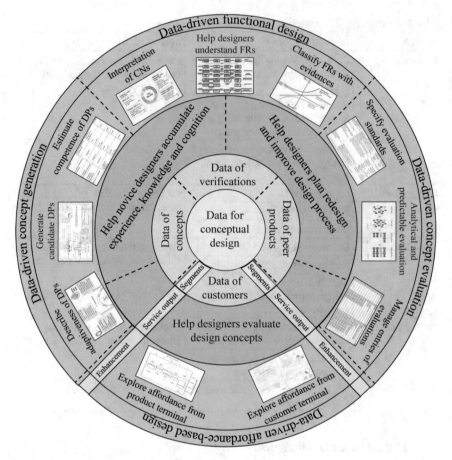

Fig. 3 Roles of data to drive conceptual design

Data analysis also helps designers explore factors (industry policies, customers' demography, product updates, etc.) that affect CNs. There are two forms of data that assist designers in generating CNs. First, intermittent data, which reflects statuses of products, environments, or customers at a specific moment, help designers understand the backgrounds of CNs. Second, consecutive data, which consist of patterns and trends of similar products and the overall market, allow designers to monitor how CNs fluctuate over time.

The collection and analysis of data are critical for the generation of FRs. FRs are based on designers' interpretation of customer voices. Data analysis of market performance and customer reviews imparts designers with a more profound comprehension of CNs to define FRs. For example, frequencies of FRs in customer voices guide designers to choose candidate FRs. Besides, designers need to rephrase FRs into the form of < verb + noun > to illustrate substances conversions (energy, material, objects, etc.) [20]. By recording data of product working, designers can trace changes of engaged substances to arrange FRs. Also, FRs are assigned with values as the target performances (e.g., Stir flour between 80 and 260 RPM) [20]. Many sources of data determine these performance values. For example, data of customer voices ensures designers understand the degree of performance customers expect. Peer products data ensure designers understand what degree of performance is necessary to remain competitive in the market. Besides, designers may generate innovative FRs to stimulate new CNs. Such innovative FRs are often from other products or based on market trends [21]. Customers' social media and online platforms are a common data source used to identify these market trends.

Data helps designers classify FRs. As introduced above, the Kano Customer Model and Long Tail Model are often applied to classify FRs. Kano Customer Model classifies FRs into excitement FRs, performance FRs, and basic FRs based on measured customer satisfaction against specified fulfilment requirements. The Long Tail Model classifies FRs based on usage popularity against the number of products of that type in the market [8, 9]. The parameters used in each model are all quantifiable. This enables designers to extract customer satisfaction levels and fulfilment requirement criteria from data sources such as customer reviews and market sales performance. Additional sources of quantitative data for fulfilment requirements include product testing and monitoring of the product when used. Data for usage popularity and number of products in the market is collected via work monitoring and market analysis of peer products. As such parameters are quantitative, the data collection and processing can be achieved autonomously by artificial intelligence. Unlike subjective data generated from a designer, such data is inherently objective.

3.4   Data-Driven Concept Generation

Data analysis allows designers to understand the adaptiveness of DPs better. When mapping DPs from CNs and FRs, designers should first analyse situations where customers are using products; for example, when a product is used, the user's behaviour, the environment the product is used in. Based on this analysis, designers should evaluate the adaptiveness of DPs to situations where the products are used. Such working situations are described by contextual data, which explain where and how a product is used. Table 2 list a series of contextual data, highlighting the situations where products are used. This includes physical contextual data, social contextual data, and user contextual data. The comparison between contextual data of working situations and contextual data of DPs helps designers select candidate DPs.

Data help designers generate candidate DPs. Designers often generate DPs by extracting and modifying existing DPs from existing design concepts or create new DPs based on existing technologies [22]. When designers extract and modify existing DPs from existing design concepts, data of peer products show the quantity of DPs for specific FRs. Additionally, data from product testing, work monitoring, and customer voices are useful for the exclusion of incompetent DPs. When designers create new

Table 2 Examples of contextual data which describe working product situations

Categories of contextual data	Example
Physical contextual data	Timetable
	Temperature
	Location
	Passenger
	Space layout
Social contextual data	Resource supply
	Notification
	Interaction with peer products
	Service output
	Entertainment
User contextual data	Demography
	Mood
	Using habits
	Preferences
	Knowledge of products
Behaviour contextual data	Energy generation
	Command
	Computation
	Movement
	Energy consumption

DPs based on enabling technologies, contextual data help designers understand the mechanisms and behaviours behind the operation of DPs.

Data help designers estimate the competence of DPs. As introduced, FRs are assigned with values of performances as design targets. Candidate DPs are designed to meet target values, with the dual aim of minimizing uncertainties and manufacturing costs. Determining DP competence is generally based on assessing how well the designers themselves understand the DPs. Therefore, this process is inherently subjective, dependent upon individual designers' knowledge and experience. The incorporation of data collection and analysis, by its objective nature, significantly enhances the assessment of DP competence. As shown in Table 2, the values of performances and attributes of DPs change under different contextual data forms. The automation and objectification of DP attribute assessment ensure the process evolves with the changing contextual data forms. Conversely, a designer's knowledge and experience are often constrained by their subjectivity and may not adapt easily to change.

Data help designers integrate DPs into design concepts. When integrating DPs into design concepts, designers must evaluate multiple candidate DPs against product functionality and compatibility with other DPs. During selections of DPs, data analysis helps designers evaluate the conformity between attributes of DPs and values of expected performance. Generally, DPs which conform gain priority. When comparing DPs, designers must consider if developing them for similar working environments, operations, and users' preferences is necessary. Contextual data promotes the segmentation and organization of DPs. DPs with similar contextual information cater to similar working environments, operations, and users' preferences.

The extraction of data from market analysis and customer voices enables new methods of design concept extrapolation. Designers tend to link DPs to popular societal topics and ideas, such as environmental protection, humanitarian issues, and technology phenomena such as the Internet. There are vast quantities of data on current trends on the Internet today, such as trending videos on social media and customer reviews on e-commerce websites. Furthermore, there are many advanced techniques for collecting and synthesizing this data. This provides more objective insights designs use for evaluating DPs or even creating new products. Besides, the extraction and analysis of peer product data provide designers with powerful insights into what other DPs have already been developed and how they can be refined.

3.5 Data-Driven Concept Evaluation

Data help designers manage entries of concept evaluations. The product development process generates an abundance of DPs. However, each DP's impact on product performance must be assessed, which requires a significant amount of designers' time. The effort involved in assessing such large numbers of DPs is significant, increasing the cost and lead time of the concept evaluation process. Therefore, to save time and effort, designers rely on their judgement and experience to evaluate what they believe to be the most vital DPs. However, this process's subjectivity may result in important DPs being overlooked or less important DPs being prioritized. A data-driven approach to DP assessment corrects many human errors common in this process. For example, data from peer products or from prior product models provides insights into how these DPs perform. Furthermore, data from customer reviews informs designers which DPs and FRs are most important to customers, and data from product usage informs designers of what features are used and which are not.

A more data-driven approach to concept evaluation also helps designers specify evaluation standards. Conventionally, designers evaluate concepts based on physical phenomenon (fracture, acceleration, computation, etc.) and customer attitudes. However, some design criteria are ambiguous and difficult to evaluate, such as assessing how artistically creative or environmentally sustainable a concept is. Consequently, the corresponding design factors for these ambiguous factors are vague and lack the specificity needed for effective concept evaluation. Designers tend to make subjective evaluations of these factors based on their own standards. In contrast, data specifies those standards into numerical form and measures concept performance.

By comparing predictions with evaluation data from physical tests, virtual simulations, and market feedbacks, designers can identify gaps between concepts and design targets. Such outcomes lead to the enhanced redesign of design concepts.

By employing a data-driven approach to conceptual design, the concept evaluation process becomes far more analytical and predictable. Design factors and engineering characteristics are initially not in a quantitative form, limiting how designers evaluate impacts of FRs or DPs and forecast potential outcomes. For example, if a beam fractures under load, designers can determine that the beam should be strengthened or the quantity increased. However, the extent to which the beam should be strengthened or how many load cycles are required before a fracture occurs demands greater precision and objectivity. Quantifying design factors and engineering characteristics is essential for their analysis and prediction. Through analysis of design evaluations, designers can investigate DP performance with respect to efficiency, failure rate, and fluctuations. Also, the input of quantified design factors and engineering characteristics into mathematical models allows for the prediction of future performance. For example, evolutionary polynomial regression uses data to model and verify dynamic environmental changes [23]; Information axiom quantifies performances of DPs and design targets via normal distribution to measure reliabilities [10].

3.6   Data-Driven Affordance-Based Design

Affordance-based design is a design process driven by customers rather than designers. Affordances are defined as non-intentional uses of designs by customers and are an unexpected phenomenon that arises from the design process. A data-driven approach is particularly useful for the investigation of affordances. Affordances are discovered via the customer or product terminals. The customer terminal involves using customer voices for the discovery of affordances. Data from the customer terminal consists of online customer discussions, shares on social media or posted videos. Analysis of this data may reveal unintended usages of products or product features which receive unexpected attention. The product terminal consists of contextual data generated from monitoring product usage. Affordances are discovered via the product terminal when real-time usage data differs from predicted results. These unexpected use cases products are then captured and investigated as potential affordances.

4   Case Study

The following section describes a real-life case study of a robot vacuum cleaner using data-driven conceptual design. RVCs are an emerging smart home product that possess many cutting-edge design features. As they are used by a wide range of customer groups, there is an abundance of data generated in both the customer and product terminals. The case study explores how such data can be collected and analysed to enhance the conceptual design process of RVCs.

4.1 Data for the Conceptual Design of Robot Vacuum Cleaner

As explained in Sect. 3.3 above, conceptual design data mainly consists of data of customers, data of concepts, data of verifications, and data of peer products. In terms of data of customers, sales of RVCs are primarily based on online platforms with an abundance of customer voice data. The demographic of customers who own RVCs are those who embrace modern technologies and social network communities. Most of them have experience using smart home products and sharing their voices online. Therefore, actual data from customer social media profiles, web browsing habits, online videos and online purchases exists in addition to product data. Several iterations of RVC designs have been developed by multiple companies, resulting in many distinct design concepts of the RVC and varieties of product data. RVCs share some common DPs with other smart home products, with examples provided below in Table 3, which provide supplementary data sources for use in the conceptual design process. Verification data of RVCs is generated from physical tests, virtual simulations, monitoring operational RVCs, and marketing channels. Additional data from peer products relating to their marketing, design features and performance is also useful for verification.

Collection of RVC product data occurs in multiple ways. As many RVCs and smart home products are sold via online platforms, with reviews posted by customers on these platforms, collecting data relating to customer voices is achievable using application programming interfaces or WebCrawler. Similarly, social media contains similar customer voice data, which is also extractable in this way. Sensors connected to RVCs, such as thermometers, ultraviolet receptors, and sonar devices, collect real-time data on how RVCs interact with their surroundings. Computer-aided design tools generate data of concepts via virtual modelling and simulations of RVC design, and the Internet of things enables RVCs to exchange data from other smart home products. Traditional methods of gathering product and customer data, such as online survey, physical tests, and studying existing concepts also play an important role in concept generation.

Table 3 Examples of data for the conceptual design of robot vacuum cleaner

Types of data	Categories of data	Examples
Data of customers	Customers' demography	78% of customers' ages are between 22 and 38; 62% of customers live in apartment; 71% of customers are single, etc.
	Customers' preferences	53% of customers prefer the sweep-mop integrated RVC; 81% of customers prefer round shape; 58% of customers prefer grey, black or white colours, etc.
	Customers' purchase history	37% of customers bought air humidifier; 73% of customers checked handheld vacuum cleaner as well; 26% of customers also searched fragrance, etc.
Data of concepts	Capability of DPs	The nominated airflow provided by motor is 65 CFM; The filtration system traps 99.9% of particles; The dirt cup capacity is 0.45L, etc.
	Popularity of DPs	71% of customers use auto-scheduled cleaning; 68% of customers download the dedicated App; 51% of customers use virtual walls, etc.
	Dimensions of geometric design	The diameter of RVC is 354 mm; The height of RVC is 105 mm; The diameter of brush is 83 mm, etc.
Data of verifications	Outcomes of physical tests	The noise generate by airflow in wind tunnel is 42 dB; The elasticity (tensile modulus) of RVC shield is 3.17GPa; The brake distance of RVC is 24-27 mm, etc.
	Outcomes of virtual simulations	Battery volume is expected to shrink 30% in 5 years; Cleaning of 180m^2 is expected to be finished in 130 min; The brush is expected to cover 97% of dirt, etc.
	Market feedbacks	32% of customers mention that battery can only support 2 h cleaning; 61% of customers mention that RVC is not satisfying when cleaning leaves; 18% of customers mention that the dedicated App is sometimes lagging, etc.

(continued)

Table 3 (continued)

Types of data	Categories of data	Examples
Data of peer products	Data of functions	25% of RVCs can mop the floor after sweeping; 80% of handheld vacuum cleaners stir debris in dirt cup; Smart home control hub support voice recognition no less than 30 dB, etc.
	Data of DPs	31% of smart home products carry wifi6 network card whilst the rest carry wifi5; The nominated airflow of handheld vacuum cleaner of the same price is 80-90CFM; 70% of RVCs are under 7 kg, etc.
	Market performance	65% of flagship RVCs of different brands are priced between $980-$1060; Brand A occupies 21% of RVC market share; Brand C has the highest customers' satisfaction rate in the previous year, etc.

4.2 Data-Driven Functional Design of a Robot Vacuum Cleaner

Demographic characteristics of customers represent an invaluable source of data for CN generation. There are many ways in which customer demographic data is collected via online platforms. For example, as RVCs are often sold via online platforms, customer data relating to gender, age, and location are recorded during the online buying process. Furthermore, users allow the online platform to access their HTTP cookies for better service, containing their purchase history and browsing data. Another unique data source comes from customer memberships to RVC brands, where customers are required to share demographic information to join. The convergence of a wide range of customer demographic data sources produces valuable insights which designers use for decision-making. For example, an insight such as '71% of customers have more than one smart home products such as smart home controller, smart music players, and smart air-conditioners', suggests most RVC customers have a range of informed experiences in smart home products. Another insight, such as '62% of customers are aged between 23 and 31', indicates that the main customer demographic is young, which may mean they have a limited financial capacity and that RVC costs should be minimized.

As most RVCs are sold online, data of customers' voices on the online platform is useful for interpreting CNs. Consider the following example of an online review left by Customer A.

'During set up of the phone app which is the centre of all the smart features. You are required to set up an account. You next are required to provide your home router password to the phone app so that the robot will be allowed to connect. The app also requires the location in the phone to be turned on. Giving this permission the app connects with the robot and I can start using it. Removing the permission crashes the app and losses the robot'.

If this post received + 1000 likes with an 83% support rate, suggesting many people have this problem, it draws attention to the fact that 7% of all complaints are about this particular problem. Therefore, the convenience of using the App is identified as an issue that needs to be optimized. A CN can be formulated as 'CN$_1$: Customers require an App with convenient usage and simple permission system'.

Meanwhile, Customer B and C comments, respectively:

'I am really happy that I only need to clean up the dust bag once a week. Every Monday I clean it up then that boy knows what it needs to do'.

'Beware of this junk. Since I used it, it either stops with 30% of the house left due to its full dust bag or gets stuck by hairs and furs'.

Both posts receive + 300 likes with a 50% supportive rate, however, each post contradicts the other. By delving into each customer's demographic data, designers find that Customer B is a male mining engineer who rents a small apartment and stays at home only during the weekend. His RVC cleans up four times a week, with little dust collected each time. Conversely, Customer C is a housewife who lives in a house and owns 2 children, 2 cats, and 1 dog. Her family members often bring debris into the house, and her pets shed hairs constantly. Therefore, her RVC cannot clean up all debris and hairs in one run, especially considering the large size of her house. Moreover, data shows that 26% of customers have troubles with dust bag's size. Amongst those customers, 64% have huge room space, 37% have more than 4 family members and 71% have one or more pets. Thus, designers can conclude a CN as 'CN$_2$: Customers have strong demands towards larger dust bags'.

Data from customer reviews are an excellent source of CNs. Customer B's comments and demography show that an RVC to clean itself automatically is important for him. If designers find that 27% customers' comments are related to the automatic cleaning function, a CN can be formulated as 'CN$_3$: Customers need the automatic cleaning of RVC'. Meanwhile, Customer C's comment indicates that her dropped hairs often impede the movement of the RVC. If designers then find that 51% of customers are female and 38% have pets, a CN can be created as 'CN$_4$: RVC should ensure smooth cleaning and prohibit being stuck by debris'.

Designers then also refer to customer voice and product data for the formulation of RVC FRs. For example, in terms of 'CN$_4$: RVC should ensure smooth cleaning and prohibit being stuck by debris', data helps designers investigate what FRs can satisfy this CN. By investigating customer voice data, designers find that 31% of complaints are about the stuck filter web in the dust bag, 26% are about the stuck wind tunnel, and the rest 43% are about the stuck wind nozzle. An investigation of physical tests and work monitoring reveals that 24% of stuck cases are caused by large debris like leaves and foams whilst the rest 76% are about flocculent or lumpy debris such as wool or hair. As a result, designers notice that splitting debris after collection is the key to avoid debris getting stuck. The formulation of two FRs is then possible: 'Stir debris during storage' and 'Split debris during transferring'. Also, designers can also extract FRs from peer products. For example, data show that

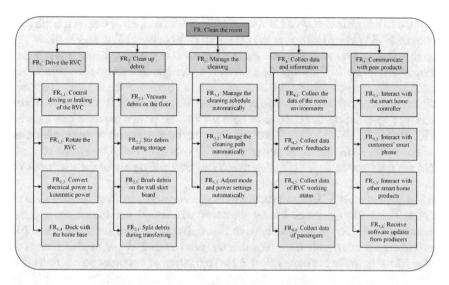

Fig. 4 Examples of functional requirements of robot vacuum cleaner

all RVCs in the market can return to charge after cleaning, then FRs of 'Dock with the home base' and 'Manage the cleaning path automatically' can be formulated. Figure 4 shows one example of an RVC's functional hierarchy.

Using data from customer voices and product operation, designers assign target values for each FR. The IDEF0 model is beneficial for collaborating data with FRs. IDEF0 presents a substance conversion to represent an FR [6]. Data quantifies input substances and output substances in this conversion, so FR's required performance can be calculated. Figure 5 shows IDEF0 models of $FR_{1.3}$ (Convert electrical power to kinematic power) and $FR_{2.1}$ (Vacuum debris on the floor) in Fig. 4. In terms of $FR_{1.3}$, assume design constraints require that the battery of RVC is 14.8 V and 2000 mA (29.6 W), the RVC is 3.8 kg and constantly moves at 0.4 m/s. $FR_{1.3}$ should provide the power to drive RVC on different materials of floors. Hypothetically, the smooth wood floor's dynamic coefficient is 0.2 whilst that of rough carpet is 0.63. Then, the power

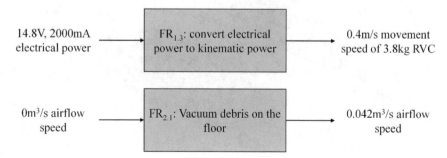

Fig. 5 Examples of IDEF0 models

required to maintain that speed is 7.46 and 23.48 W. Thus, $FR_{1.3}$ can be specified as 'convert 29.6 W electrical power to kinematic power between 7.46 and 23.48 W'. In terms of $FR_{2.1}$, assume design constraints require that the maximum airflow speed allowed is $0.042\,\text{m}^3/\text{s}$. Knowing the air density under standard atmosphere pressure is $1.225\,\text{kg/m}^3$, drag coefficient is 0.217, and area of nozzle surface is $1.26 \times 10^{-9}\,\text{m}^2$, it then can be calculated that the power required to vacuum is 5.71 W. The efficiency of power transferring is assumed to be 35–45%, the power required can then be calculated as 12.69–16.31 W. Therefore, $FR_{2.1}$ can be specified as 'vacuum debris on the floor with 12.69–16.31 W power output'. It should be noted that some of these data are the hypothesis for the illustration and may not represent the real work of RVC.

Using data from CNs and product testing, designers can then classify FRs of RVCs. The Kano Customer Model measures customers' satisfaction against the performance of FRs, as shown above in Fig. 6. Both customer satisfaction and FR performance are derivable from customer and product data. For example, in terms of $FR_{1.1}$ (drive or brake the movement), an investigation of customer voices finds that 76% of customers complain that the braking systems are incompetent, resulting in RVCs damaging their furniture. 24% of customers complained that the RVC moves too slower than they expect. Importantly, few customers expressed positive opinions about the RVC's braking system. Therefore, designers conclude that $FR_{1.1}$ is a basic

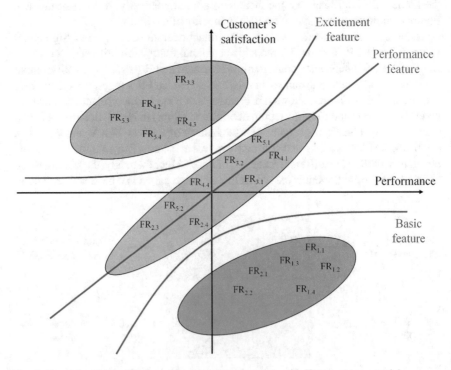

Fig. 6 Examples of classification of functional requirements with Kano customer model

feature as customers have no tolerance to its problems whilst it is indispensable. In terms of $FR_{4.2}$ (collect data of users' feedbacks), an investigation of customer voices finds that 47% of customers enjoy the convenience of this function, 42% mentioned that the feedback captured is inaccurate, and the remaining 11% preferred conventional feedback methods of posting on the website. Meanwhile, designers find that 27% of customers have never used this FR and only 3.4% use it every time after cleaning. Thus, $FR_{4.2}$ is classified as an exciting feature as it (1) improves customers' overall satisfaction; (2) customers do not emphasize its performance; (3) its popularity is low. Figure 6 shows the classifications of FRs based on the Kano Customer Model.

4.3   Data-Driven Concept Generation of Robot Vacuum Cleaner

Designers firstly generate candidate DPs, data of previous concepts, and peer products of RVCs help designers extract existing DPs. For example, in terms of $FR_{4.4}$, it can be specified as 'collect data of passengers at a range between 0.5 and 1.7 m within 200 ms'. By investigating previous concepts, data show that 57% RVCs detect obstructions and passengers via shock-absorbing bumpers, 28% detect with optical emitter, and the rest 15% detect with cameras. Thus, the shock-absorbing bumper, optical emitter and cameras can be three candidate DPs of $FR_{4.4}$. Meanwhile, when cloud sourcing 'remote detection of obstruction', designers find that 17% results are related to 'Internet of things' and 11% are related to 'surveillance camera'. Designers then notice that the two keywords in combination lead to the monitor network in logistic warehouses. In those logistic warehouses, the monitor network assembled by surveillance cameras captures the real-time layout of the whole warehouse and transfers to each transport robot. Those transport robots thus know their distances to obstructions and avoid in time. Therefore, 'monitor network connected to RVC' can be a candidate DP as well. Table 4 shows some examples of DPs of the RVC. Besides, data of concepts help designers estimate competence of DPs. For example, the shock-absorber bumper can only detect objects when there is more than 0.3 mm deformation, which conflicts to the required 0.5–1.7 m distance of detection. Therefore, designers can exclude shock-absorbing bumper without design evaluation.

Contextual data help designers understand the adaptiveness of DPs. Using the example of $FR_{4.4}$, optical emitters, cameras, and the monitor network are formulated as candidate DPs. Table 5 shows some examples of contextual data describing the working situation of RVCs (the percentage value in the table represents the proportion or popularity). Designers use contextual data to determine if the candidate DPs align with the working situation of RVCs. Optical emitters and cameras do not conflict with the working situation contextual data. However, for the monitor network, the fundamental of monitor network is the Internet of things of multiple surveillance cameras in users' home. Contextual data of the working situation shows that only

Table 4 Examples of design parameters of robot vacuum cleaner

Functional requirements	Design parameter 1	Design parameter 2	Design parameter 3
$FR_{1.1}$	Feedback control system	Logic control system	Open-loop control system
$FR_{1.2}$	Four-wheel chassis with two rotatable front wheels	Three-wheel chassis with all wheels fully rotatable	Chassis with tentacles to move
$FR_{1.3}$	Stepper motor	Servo motor	DC brushless motor
$FR_{1.4}$	Card slot design of base	Flat design of base with wireless charging	Magnetic docking
$FR_{2.1}$	Flow-through motor	Peripheral bypass motor	Tangential bypass motor
$FR_{2.2}$	Stirring machine	Motor-blade design	Air flow
…	…	…	…

9.7% of users have surveillance cameras at home (including those with only one at home). Therefore, the monitor network is not applicable for most users and thus should be excluded from candidate DPs. Similarly, designers can use contextual data to compose DPs into ICs. Those DPs (for different FRs) with similar contextual data can be composed together whilst those with significantly different contextual data will exclude each other.

4.4   Data-Driven Concept Evaluation of Robot Vacuum Cleaner

Data of historical concept evaluations remains a salient source of information for new versions, as they inform designers of tried and tested approaches from other designers. The customer voice is a central tenet of the concept evaluation process. For example, hypothetically, designers set a threshold for entries worthy of evaluations: mentioned in over 80% of historical cases or 3% of all customer voices. In this condition, 'Endurance of buttons on control panel' is only mentioned by 0.84% customer voices but evaluated 93.4% in previous cases. Therefore, evaluating 'endurance of buttons on control panel' should be prioritized. 'Sensitivity of infrared sensors' is excluded because its existence rates in previous evaluation cases and customer voices are only 71.4% and 1.6%, respectively. Table 6 shows some examples of entries of concept evaluations. Designers can adjust the threshold according to available time and costs.

Once designers have completed an evaluation of prior concept evaluation methods, they must specify standards and evaluate DPs. The Information Axiom [5] provides an effective approach to achieving both. Using $FR_{1.3}$ (convert 29.6 W electrical power to kinematic power between 7.46 and 23.48 W) as an example, designers will evaluate the maximum kinematic power (23.48 W) of corresponding DPs (stepper motor, servo

Table 5 Contextual data examples of working situations of robot vacuum cleaner

Contextual data	Working situation of robot vacuum cleaner
Timetable	17% 8am-10am, 14.2% 10am-12 pm, 9.7% 12–2 pm
Temperature	14.1% 8–14 °C, 12.7% 14–17 °C, 28% 17–22 °C, etc.
Location	71% indoors, 29% outdoors; 87% dry, 13% wet
Passenger	41% adults, 17% teenagers, 11% elders, 14% cats, 17% dogs
Space layout	24% with mess sundries, 81% with more than 8 plants or furniture, 1.7% completely empty, etc.
Resource supply	/
Notification	13% of cleaning reminding, 34.1% of cleaning status, 4.2% of software update, 12.4% of advertisement, etc.
Interaction with peer products	17% with smart home controller, 41% with smartphone, 14% for command output, 27% for data exchange, etc.
Service output	41% of debris suction, 12% of floor mopping, 24% of data supply, etc.
Entertainment	/
Demography	41% are between 22 and 26 years old, 61.6% receive tertiary education or higher, 57.7% are single, etc.
Mood	During cleaning, 68% users feel normal, 7.1% feel excited, 4.7% feel annoyed, etc.
Using habits	14% clean once a week, 17% clean every day, 24% never use RVC to clean kitchens, etc.
Preferences	17% reject square shape design, 41% prefer polymers than others, 22% reject white, grey, or black colour, etc.
Knowledge of products	71% have sufficient experience in using smart home products, 9.7% have surveillance cameras at home, 57% used RVC before, 9.8% have abundant knowledge about RVC, etc.
Energy generation	5–36 V voltage generated, 0.017–0.043m^3/s air flow, 14.63–22.94 W kinematic power, etc.
Command	/
Computation	44% are about cleaning management, 37% are about operation management, 19% are about forecast, etc.
Movement	7.6% acceleration, 6.8% braking, 79% linear movement, 13.7% rotation, etc.
Energy consumption	47% are dissipated by heat, 42% are transferred to air and debris, 61% RVCs consume more than 0.63KWh for every cleaning, etc.

motor, and DC brushless motor). Figure 7 presents the information axiom for $FR_{1.3}$. Next, designers specify that the design target is 23.48 W and the tolerance is 5% (22.31–24.65 W). Considering the power output turbulence, designers then specify that the power output should not be less than 21.5 W or more than 25 W during operation. Stepper motor A, servo motor B, and DC brushless motor C are selected for further evaluation based on these standards. The maximum power of each motor

Table 6 Examples of entries of concept evaluations

Evaluation entries	Percentage of evaluation in previous cases (%)	Percentage of existence in customer voices (%)
Endurance of battery	97.4	14.3
Suction power	96.3	8.6
Data transferring speed	82.7	2.7
Accuracy of navigation	92.5	8.4
Split effect in wind tunnel	80.4	1.3
Overheat protection	88.2	0.7
…	…	…

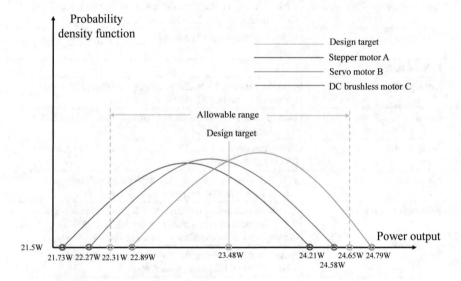

Fig. 7 Example of information axiom for $FR_{1.3}$

is plotted as a probability density function in Fig. 7. Designers then determine each motor's reliability via calculation of the cross-sectional area between the allowable range and each probability density function.

4.5   *Data-Driven Affordance-Based Design of Robot Vacuum Cleaner*

Affordances of RVCs are found by investigating unexpected findings in data of product operation and customer. For example, an interesting existing affordance of RVC is that some users use RVCs to carry their pets (especially cats) for entertainment.

Users often post photos or videos on their social media or online video platforms from the customer terminal, such as 'cats vs robot vacuum', 'Roomba cats' and 'cat logic on a robot'. During RVC development, the expected relationship between a user's pet and the RVC is focused on cleaning the pet's fur and avoiding contact with the pet. Notably, using data collected by web crawlers, social monitoring, or crowdsourcing from customer social media accounts uncovers this affordance. Finding the CN of entertaining pets, it is also possible to formulate this as an FR.

Data collected during the RVC's operation in the product terminal is also invaluable for discovering affordances. Generally, from prior testing and feedback from previous products, designers have quantified values for how customers will use the product. Unexpected uses of the RVC, such as entertaining a user's pet, also generate unique sensor data from the RVCs. For example, some videos online show RVCs carrying cats around rooms. Load sensors will detect higher than the usual load on the upper surface of RVC, enabling designers to identify this affordance.

5   Conclusion

In conclusion, this chapter introduces the data-driven conceptual design. Data-driven conceptual design provides three services: (1) Helping novice designers accumulate experience, knowledge, and cognition; (2) Helping designers redesign and improve the design process; (3) Helping designers verify design concepts. Data-driven conceptual design significantly benefits each of the four stages of the design process. Firstly, it helps designers interpret CNs, understand FRs, and classify FRs with evidence in functional design. Secondly, a data-driven approach highlights the adaptiveness of DPs, generates candidate DPs and estimates the competence of DPs in concept generation. Thirdly, it promotes evaluation entries, specifies evaluation standards, and facilitates more accurate analysis and predictions. Fourthly, it leads to discovering affordances from both product and customer terminals. Finally, in contrast to conventional empirical, conceptual design, a data-driven approach reduces the heavy reliance on designer knowledge, making the design process more objective, analytical, and replicable.

References

1. Dym, C. L., Little, P., Orwin, E. J., & Spjut, E. (2009). *Engineering design: A project-based introduction.* John Wiley and sons.
2. Umeda, Y., Ishii, M., Yoshioka, M., Shimomura, Y., & Tomiyama, T. (1996). Supporting conceptual design based on the function-behavior-state modeller. *Ai Edam, 10*(4). 275–288
3. French, M. J., Gravdahl, J. T., & French, M. J. (1985). *Conceptual design for engineers.* London: Design Council.

4. Wu, D., Rosen, D. W., Wang, L., & Schaefer, D. (2015). Cloud-based design and manufacturing: A new paradigm in digital manufacturing and design innovation. *Computer-Aided Design, 59*, 1–14.
5. Suh, H. P. (2001). *Axiomatic design: Advances and applications MIT-Pappalardo series in mechanical engineering.* USA: Oxford University Press.
6. Kim, S. H., & Jang, K. J. (2002). Designing performance analysis and IDEF0 for enterprise modelling in BPR. *International Journal of Production Economics, 76*(2), 121–133.
7. Liu, A., Wang, Y., & Dai, Y. (2018). Customer review-driven functional formulation for design education. In *2018 ASEE Annual Conference & Exposition.* American Society for Engineering Education.
8. Chen, C. C., & Chuang, M. C. (2008). Integrating the Kano model into a robust design approach to enhance customer satisfaction with product design. *International Journal of Production Economics, 114*(2), 667–681.
9. Anderson, C. (2006). The long tail: Why the future of business is selling less of more. Hachette Books.
10. Suh, N. P. (1998). Axiomatic design theory for systems. *Research in Engineering Design, 10*(4), 189–209.
11. Huang, G. Q., & Mak, K. L. (1999). Web-based morphological charts for concept design in collaborative product development. *Journal of Intelligent Manufacturing, 10*(3–4), 267–278.
12. Bryant, C. R., McAdams, D. A., Stone, R. B., Kurtoglu, T., & Campbell, M. I. (2005). A computational technique for concept generation. In *ASME 2005 International Design Engineering Technical Conferences and Computers and Data in Engineering Conference* (pp. 267–276). American Society of Mechanical Engineers.
13. Holtzblatt, K., & Beyer, H. (1999, May). Contextual design: Using customer work models to drive systems design. In *CHI'99 Extended Abstracts on Human Factors in Computing Systems* (pp. 139–140). ACM.
14. Maier, J. R., & Fadel, G. M. (2009). Affordance based design: A relational theory for design. *Research in Engineering Design, 20*(1), 13–27.
15. Ciavola, B. T., & Gershenson, J. K. (2016). Affordance theory for engineering design. *Research in Engineering Design, 27*(3), 251–263.
16. Kirby, M. R. (2001). A methodology for technology identification, evaluation, and selection in conceptual and preliminary aircraft design. Doctoral dissertation, Georgia Institute of Technology.
17. Levin, J. A., & Datnow, A. (2012). The principal role in data-driven decision making: Using case-study data to develop multi-mediator models of educational reform. *School Effectiveness and School Improvement, 23*(2), 179–201.
18. Liu, A., Wang, Y., & Dai, Y. (2018). Customer review-driven function formulation for design education. In *2018 ASEE Annual Conference & Exposition.*
19. Kumar, K. S., Desai, J., & Majumdar, J. (2016). Opinion mining and sentiment analysis on online customer review. In *2016 IEEE International Conference on Computational Intelligence and Computing Research (ICCIC)* (pp. 1–4). IEEE.
20. Shah, J. J., & Rogers, M. T. (1988). Functional requirements and conceptual design of the feature-based modelling system. *Computer-Aided Engineering Journal, 5*(1), 9–15.
21. Johansson-, U., Woodilla, J., & Çetinkaya, M. (2013). Design thinking: Past, present and possible futures. *Creativity and Innovation Management, 22*(2), 121–146.
22. Liu, A., & Lu, S. C. Y. (2015). A new coevolution process for conceptual design. *CIRP Annals, 64*(1), 153–156.
23. Giustolisi, O., & Savic, D. A. (2009). Advances in data-driven analyses and modelling using EPR-MOGA. *Journal of Hydroinformatics, 11*(3–4), 225–236.

Chapter 4
Management of Constraints, Complexities, and Contradictions in the Data Era

Abstract Rapid product developments and market changes significantly impact how designers manage concept improvements in conceptual design. Data collection and analysis techniques are becoming critical tools for managing the increasingly complex concept improvement process. The advantages of the data-driven approach lie in its objectivity, as traditionally, design decisions depend on designers' subjective opinions. This chapter introduces a data-driven approach for managing constraints, complexities, and contradictions. Firstly, the theory behind constraints, complexities, and contradictions is clarified. Secondly, the three primary benefits of a data-driven approach to concept improvement are explained, including (1) How data helps designers fathom constraints, complexities, and contradictions; (2) How data delineates the influence of constraints, complexities, and contradictions on FRs; (3) How data is used to specify concept improvement design targets. These benefits are explained through constraint, complexity, and contradiction management methodologies. Finally, a case study of a robot vacuum cleaner illustrates this data-driven approach to the concept improvement.

Keywords Conceptual design · Concept improvements · Smart design

1 Introduction

Conceptual design consists of concept generation and concept improvement (CI) processes. Concept generation includes functional design, concept evaluation, and affordance-based design (introduced in the previous chapter). The CI process explained in this chapter uses theories and practices of constraint management (CTM), complexity management (CPM), and contradiction management (CDM) for enhanced design decision-making.

CIs are an integral part of conceptual design, as the constraints, complexities, and contradictions of design concepts significantly impact their performance and risk of failures. Constraints are facets of a design that restrict functional requirements (FRs) and design parameters (DPs) [1, 2]. Complexities are unforeseen barriers to FR executions and potential failures in the design concept [3]. Contradictions limit the design configuration feasibility and the realization of DPs [4]. Conventionally, designers regard CIs as objectives for solving problems caused by constraints, complexities, and contradictions [5, 6]. Scholars are increasingly noticing the value of ideation behind constraints, complexities, contradictions, and the design processes that spur innovation [7–9].

CIs are regarded as some of the most challenging design activities. Similar to other conceptual design procedures, CIs collaborate subjectivities and objectivities [10–12]. Subjectivities relate to the recognition, compilation, and management of constraints, complexities, and contradictions determined by designers' knowledge, cognition, and experience. In comparison, objectivities involve the existence and influence of constraints, complexities, design configurations, and physical environments based on contradictions [11, 12]. In concept generation, the intangibility of constraints, complexities, and contradictions makes their description and analysis difficult for designers. CI creation is characterized as a highly iterative design process. Design concepts are continuously modified and compromised based on constraints, complexities, and contradictions. This process is described as a 'dynamic loop' where designers repeatedly challenge their own concepts.

Data-driven approaches for concept improvement are becoming progressively vital for conceptual design. Advances in computer-aided design in recent decades have transformed design factors into digital data forms. Consequently, the importance of using data for calculation, modelling, and verification in engineering design has burgeoned [13]. The emerging trends of smart manufacturing and product digitalization mean data collection and analysis will become the foundation of each manufacturing process. This includes design, fabrication, maintenance, and the improvement of management, technical support, and collaboration [14]. Despite the intangibility of constraints, complexities, and contradictions, it is possible to collect data related to such factors and model their impact on design concepts. Furthermore, data analysis is required for setting design targets and monitoring the effect of concept improvements. A data-driven design process will shift the conventional, empirical method of assessing concept improvements into a more analytical, predictable, and controllable process.

This chapter explores the idea of how a data-driven approach impacts concept improvement in the following sections. First, Sect. 2 will introduce theory backgrounds of constraints, complexities, and contradictions. Second, Sect. 3 will introduce the challenges of empirical CIs and the importance of data. Third, Sect. 4 will introduce data-driven concept improvements. Fourth, Sect. 5 will present a case study of the robot vacuum cleaner to illustrate proposed theories. Finally, Sect. 6 will conclude this chapter.

2 Theory Backgrounds of Constraints, Complexities, and Contradictions

2.1 Introduction to Constraints

In conceptual design, constraints are design factors that influence the concept to achieve an ideal function or state. During the task clarification phase, constraints introduce restrictions to potential design solutions [15]. For example, signal strength is a constraint of a mobile phone that influences its primary function of making and receiving calls. Boundaries are an essential element of constraints, as they restrict specific value ranges of FRs and prescribe design specifications. Exceeding boundaries result in functional or structural failures of products. For example, the maximum shear stress load of a motor shaft is 15 MPa. Therefore, 15 MPa is the upper boundary for the corresponding output power. As many FRs and constraints are both interpreted from customer voices, it is not rare to see novice designers confuse the two design methods [16]. For example, constraints such as 'minimize weight', 'minimize cost', and 'maximize strength' are often interpreted as FRs by novice designers.

According to Liu's theory, based on sources, constraints can be classified into four categories: input internal constraints, system internal constraints, input external constraints, and system external constraints. Input internal constraints contain those deliberately input from stakeholders or systems for mandatory requirements. For example, 'non-toxic' is a constraint introduced by regulatory authorities for all home product designs. The implementation of such constraints is compulsory for all design concepts. System internal constraints are those introduced by other product parts. Some of these constraints are deliberately inputted by designers to control coupled parts, whilst others unintentionally affect each other. For example, in a smartphone, the 'power-saving mode' is a constraint introduced by designers, restricting computation power and signal strength to extend battery life. Input external constraints are introduced by a product's working environment, such as environmental factors that directly impact FRs. For example, the operating environment temperature is a constraint that affects battery working of electric cars. System external constraints are introduced by designers' subjectivities and are an inherent part of the design process. For example, if environmental protection is an intended theme of the design, a constraint for material selection may be 'recyclability' [7].

CTM is a process of concept compromises and concept ideations. Concept compromises occur when modifying DPs to fall within constraints are only achievable at the expense of other areas of a concept's performance. Here designers must prioritize some features over others and achieve an optimal compromise between product functions. In contrast, concept ideations occur when modifying DPs improves the overall performance of concepts without the need to compromise some features over others. For example, an aircraft longeron's weight is constrained below 66 lbs but exceeds the current longeron design. A concept compromise approach involves changing the present steel alloy into an aluminium alloy, a lighter material but with reduced strength. In comparison, a concept ideation approach requires a redesign of the longeron geometry or inventing a lighter material with higher strength.

2.2 Introduction to Complexities

In conceptual design, there are two standard definitions of complexities. In the physical domain, complexities are defined as the information required to describe the state of a system [3]. In the functional domain, complexities are defined as uncertainties to achieve FRs [17]. Both definitions concur that complexities influence decision-making processes, failure risks, and manufacturing costs of concepts. These complexities are, by their nature, measurable. In the physical domain, according to Frizelle and Woodcock's theory, complexities can be measured by system entropy:

$$H_{static}(S) = -\sum_{i=1}^{M}\sum_{j=1}^{N_j} p_{ij}\log_2 P_{ij} \tag{1}$$

$$H_{dynamic}(S) = -Plog_2 P - (1-P)Plog_2(1-P) - (1-P)$$
$$\times \left(\sum_{i=1}^{M^q}\sum_{j=1}^{N^q} p_{ij}^q\log_2 P_{ij}^q + \sum_{i=1}^{M^m}\sum_{j=1}^{N^m} p_{ij}^m\log_2 P_{ij}^m + \sum_{i=1}^{M^b}\sum_{j=1}^{N^b} p_{ij}^b\log_2 P_{ij}^b\right) \tag{2}$$

where $H_{static}(S)$ and $H_{dynamic}(S)$ represent the entropy of a static system and a dynamic system, respectively. In Formula (1), M represents the number of resources, N_j represents the number of possible states at resource j. P_{ij} represents the probability of resource j =in state i. In Formula (2), in addition to Formula (1), P represents the probability that the system is under control, P^q represents queues with length longer than 1, P^m represents queues with a length between 0 and 1, P^b represents the non-programmable states [18]. In the functional domain, Suh's theory measures complexities by effective information content:

$$I = \log_2\left(\frac{R_s}{R_c}\right) \tag{3}$$

As shown in Fig. 1, the system range is defined by performances and capability to achieve corresponding functional requirements. Expected functional requirements and design tolerances define the design range. R_s represents the range defined by system probability density, which is equal to 1; R_c represents the common range between system range and design range; The information content I is the logarithm of the ratio between R_s and R_c with a base of 2 [17].

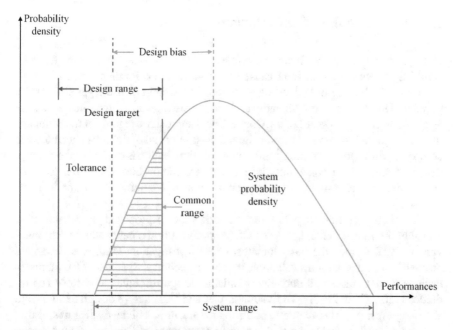

Fig. 1 Measurement of information content

Complexities are classified according to how they change over time. Constant complexities are classified as static complexities in the physical domain or time-independent complexities in the functional domain. An example of static complexity is the number of product parts, which typically does not change once a product has been manufactured and therefore does not change over time. In comparison, complexities that change over time are classified as dynamic complexities (or time-independent complexities) in the physical domain or time-dependent complexities in the functional domain. An example of a dynamic complexity are customer's buying habits, which typically change over time as social trends and product offerings evolve.

CPM is a concept modification process for reducing complexities. Fewer complexities ensure concepts are more reliable and robust. Designers evaluate existing complexities by calculating uncertainties or system entropy. If complexities exceed acceptable ranges, designers must modify concepts to reduce them. According to Axiomatic Design Theory, reduction of complexities is achieved by removing couplings among FRs and DPs. Meanwhile, some scholars argue that designers must embrace complexities for promoting product innovation and novelty [19].

2.3 Introduction to Contradictions

Contradictions represent incompatibilities between concepts' design features. Conflicting FRs are an inherent cause of design incompatibilities. For example, a chair must be both light and strong to support the weight of users yet still remain portable. However, as strength and weight are positively correlated, the FR 'strong and light' becomes a contradiction. Conventional approaches regard contradictions as inherent system problems, where designers must select an optimal balance between competing FRs. However, each contradicting design feature is critical to the concept design, distinct from constraints where they are removable. Therefore, to avoid compromises, it is common for designers to remove contradictions from concepts entirely.

The Theory of Inventive Problem-Solving (TRIZ) proposed by Altshuller is arguably the most vital and adopted CDM approach in the pattern of system innovation. TRIZ formed the basis for most CDM investigationsTRIZ was developed through investigations on evaluation trends on technical systems. TRIZ aims to exclude compromises and maximize available design resources to remove contradictions [20]. Theoretically, TRIZ summarizes and classifies 48 design features into six categories: physical features, performance features, efficiency features, utility features, manufacturing features, and measurement features. Contradictions between design features are organized in a 48 × 48 TRIZ Matrix and classified into technical and physical categories. TRIZ proposes 40 elimination principles to remove technical contradictions and four principles to solve physical contradictions. Each technical contradiction has its recommended resolution principles shown in a TRIZ Matrix. To apply TRIZ, designers must first identify design problems in concepts. If the contradiction is a technical category, designers match engaged design features in a TRIZ Matrix to find recommended principles. If the contradiction is a physical category, designers seek solutions from principles of supersystem, subsystems, separation of time, and separation of space [21].

3 Importance of Data for Managing Constraints, Complexities, and Contradictions

3.1 Challenges of Concept Improvements

As stated above, CIs significantly challenge designers' empirical thinking due to the intangibility, ambiguity, and subjectivity of constraints, complexities, and contradictions. Novice designers often struggle with the CI process due to their lack of experience, knowledge, and cognition. As constraints, complexities, and contradictions are qualitative design factors, describing them in precise engineering language is difficult. As a result, despite the necessity of CIs for achieving FRs, comprehending their effect on FRs, specific design targets and how many optimizations are needed remains a constant challenge for designers. Designers often confuse the intent of constraints, complexities, and contradictions as they intersect within similar design features and overlap with some FRs. Hence, designers are prone to select sub-optimal CI strategies.

CIs are characterized as highly iterative design procedures. When designers identify design problems caused by constraints, complexities, and contradictions, subsequent CI improvement strategies often introduce new design problems. Therefore, few design cases are ever complete without multiple CI iterations. As mentioned earlier, constraints, complexities, and contradictions are concept-based factors, where modifications to concepts change their intents. Therefore, CIs constantly challenge designers' comprehension of a design's features and problems. Today, market competitiveness is compressing conceptual design lead times, forcing designers to develop competent CIs within a limited time.

The development of more sophisticated product designs significantly increases the occurrence of constraints, complexities, and contradictions. Product developments are shifting from a monodisciplinary mode to a multidisciplinary mode. For example, traditional mechanical products such as washing machines and vehicles now contain an abundance of electronics and computer algorithms. Products also possess greater functionality and personalization mechanisms, further expanding the number and complexity of product components. The quantity of constraints, complexities, and contradictions is growing exponentially in modern-day conceptual designs, requiring more a productive CI approach.

3.2 Roles of Data in Concept Improvements

Data collection and analysis of product performance and customer voices help develop constraints, complexities, and contradictions. Many elements of constraints, complexities, and contradictions are describable in both qualitative and quantitative data forms. For example, although constraints are characterized as intangible design factors, they are determined by quantitative boundaries. Data also provides objective descriptions of the conditions where constraints, complexities, and contradictions occur. For example, as mentioned in Chap. 3, contextual data is useful when specifying a product's working situation, as it describes how a product is used.

Data delineate influences of constraints, complexities, and contradictions on FRs. In most design cases, customers do not equivalently concentrate on different design features. Besides, design resources are limited. Therefore, designers must evaluate the influence of constraints, complexities, and contradictions on FRs and prioritize them accordingly. Analysis of product data identifies correlations among constraints, complexities, contradictions, and FRs. Plotting these correlations in graphical forms enables designers to monitor variances of constraints, complexities, and contradiction data and make more informed decisions.

Data specify design targets of CIs. Conventionally, designers used their subjective judgments to determine a CI's competence. However, designers' own biases are sometimes inconsistent with technical requirements and customer needs. Product and customer data analysis aims to reduce designers' own biases and the ambiguity of this process. For example, in CDM, if customers customized a mobile phone with a minimum display size of 6.2 inches and a maximum weight of 5.5 oz, designers' CDM solution must meet these two targets.

4 Data-Driven Concept Improvements

4.1 Data-Driven Constraint Management

Figure 2 present the roles of data in constraint management. As shown, data describes the various contexts of design constraints and their dynamic relationship with concepts and environments. For example, when designing acoustic systems for air conditioners in a university, the classroom's noise level must be constrained below 4 dB, and below 20 dB in the library. Any design strategy must incorporate specific scenarios to remain feasible. Conventionally, awareness of such scenarios relied on designers' empirical thinking. However, increasingly sophisticated product designs and constraints are fast surpassing even experienced designers' capabilities. Therefore, the application of advanced data collection and analytics is required for designers to comprehend product environments accurately. Many environmental factors are measurable by both qualitative and quantitative data forms. For example, as described in Chap. 3, contextual data provides information about a design concept's working environment. Analysis of contextual data allows designers to compare how products perform in different design situations. Such contextual data includes environmental temperature, operational speed, humidity, weight and uses by different

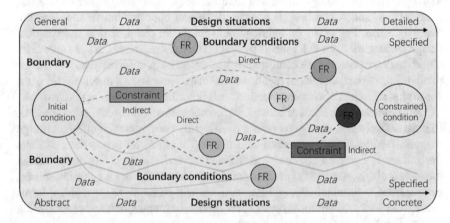

Fig. 2 Roles of data in constraint management

customer demographics—all of which are dynamic and variable. When testing products, collecting and processing the most critical environmental data presents useful information to help designers assess and predict the impact of constraints on product designs.

Data specify boundary conditions of constraints. Boundary conditions are arguably the most important constraint property, as they define ranges for what must and what cannot be achieved. In terms of design ideations, boundary conditions guide generations of new concepts to suit constraints. When specifying boundary conditions, data have two primary goals. The first is to specify the ranges of permissible solutions retrained by constraint boundaries. The second is to describe the influence on FRs when constraint boundaries are crossed. A variety of influences on FRs determine constraints and affect strategies of CTM.

Data indicate couplings among constraint-FR and constraint-constraint. Constraints directly influence on FRs with varieties. Data indicate reactions of FRs subject to changes of constraints. For example, Wi-Fi strength is a constraint that directly affects a mobile phone's browsing speed. When above the lower boundary, data shows how Wi-Fi strength promotes browsing speed in a parabolic line. If Wi-Fi strength crosses the lower boundary, the browsing speed drops significantly. The influence of some constraints on others also indirectly influences FRs. For example, using the same example of a mobile phone's Wi-Fi browsing speed, the environmental temperature in which the phone is used does not directly affect browsing speed. However, high temperatures can trigger the component protection system that automatically reduces a mobile phone's computation power. Hence, environmental temperature is an indirect constraint on browsing speed. As both the environmental temperature and Wi-Fi speed are quantifiable, it is possible to measure and identify the exact point at which the former parameter affects the latter.

4.2 Data-Driven Complexity Management

Data measure the complexities of design systems. Conventional CPM approaches were highly qualitative. However, new data-driven, quantitative methods are far more superior for design process control and system behaviour prediction [3]. Based on Frizelle and Woodcock's research on entropy theories and Suh's work on information theories, complexities are measurable in both the physical and functional domains [17, 18]. Frizelle and Woodcock's theory focuses more on operational complexities, whilst Suh's theory focuses on FR uncertainties. Probability terms are the common paradigm designer's use for measuring and evaluating complexities, whilst outcomes are formalized in information terms [3]. This chapter will use Suh's Axiomatic Design Theory for clarifications. Based on Axiomatic Design Theory, Fig. 3 illustrates the role of data in complexity management. Measurement of complexities by Axiomatic Design Theory begins with designers arranging DP couplings to achieve an FR. Designers begin with arranging couplings of DPs to achieve a FR. Then, for the work of each DP, designers signify the design target, the tolerance, and the system

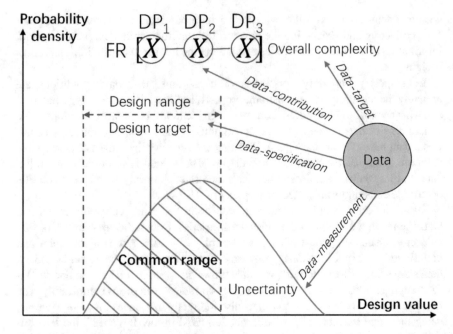

Fig. 3 Roles of data in complexity management

range to calculate the common range. Using Formula (3), designers determine the information content of each DP. Multiplying information contents of coupled DPs produces the overall information content to achieve the FR. A higher information content quantitatively represents more efforts to achieve the FR.

When analysing data for CPM target specification, scholars often apply a 'minimalistic' approach, which means continuously reducing complexities wherever possible. However, in practice, CPM is restricted by limited time, stakeholders' requirements, available resources, and designers' capacity for problem-solving. In some cases, designers deliberately reserve complexities for promoting novel concepts. Therefore, a clear target of CPM is necessary for balancing these design issues. Data-specified design targets are more specific and objective than targets derived from designers' own empirical decisions. Specific and quantified data allows for the efficient analysis of design concept deficiencies and the anticipation of improvements achieved via CPM. Furthermore, quantitative targets provide greater transparency of the effort required for CPM and other design strategies.

Data indicate contributions of complexities from each source. The fundamental goal of CPM is removing complexities based on design targets. System complexities comprise multiple design factors such as couplings of FRs-DPs, working turbulence of DPs, and environmental factors. Conventionally, the removal of complexities is subjective and random, based solely on a designer's thinking. As a result, CPM outcomes are often imbalanced, where some complexities are removed, whilst others remain unchanged or overshot. Collecting data from each design factor enables the contribution of their complexities to be measured. Then, complexities are removed through redesigning FRs, changing DPs, and modifying design features. This removal process must achieve a balance between concept reliability and design effort required.

4.3 Data-Driven Contradiction Management

As shown in Fig. 4, data have three important roles in CDM procedures. The first role is to clarify the contradictions which exist between design features. Designers are often challenged to describe how two significant design features are incompatible with each other. Enabled by data, contradictions can be numerically elucidated to make the effect of one design feature plotted against each other. For example, smartphone users expect the battery to be portable and long lasting. However, these two design features are contradictory, a greater storage capacity generally requires a larger battery size. Based on product data, it is possible for designers to model how compressing the battery size impacts its capacity. Such numerical clarifications of contradictions are essential when designers adopt the conventional CDM approach. It enables designers to decide what features they must compromise and by how much.

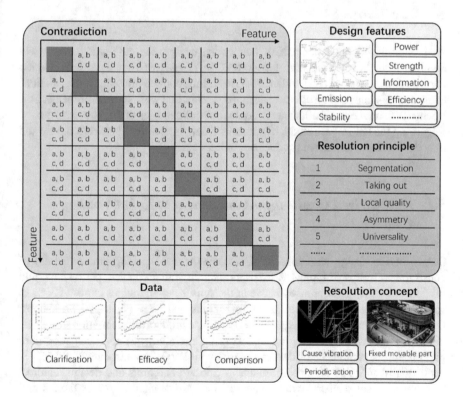

Fig. 4 Roles of data in contradiction management

Data help designers evaluate the efficacy of CDM The intangibility of contradictions and few governing standards limit designers' ability to scrutinize their CDM solutions. Including more quantitative product and customer data more accurately reflects the competence of CDM. When contradiction compromises occur, designers will compare the values of two design features throughout CDM. Designers often compare design features with FRs to evaluate the significance of contradictions in CDM. In terms of contradiction removals, data present improvements of design features throughout CDM. Also, data are important to evaluate additional design resources used for contradiction removals.

It is also possible to use data to compare different CDM strategies. As shown in the TRIZ Matrix in Fig. 4, multiple alternative principles resolve each contradiction, with infinite solutions under each principle. Despite the non-existence of a perfect design solution, designers select an appropriate resolution concept for the specific design case. Data enable measurements of design factors related to CDM such as performances of FRs, uncertainties of DPs, and manufacturing difficulties. Comparisons of CDM solutions of CDM highlight numerical differences among design factors, helping designers select candidate solutions.

5   Case Study

As CIs comprise the latter half of conceptual design, the case study below will continue from the robot vacuum cleaner (RVC) examples used in the previous chapter. RVC design consists of an abundance of constraints, complexities, and contradictions caused by designers' subjectivities, design concepts, and working environments. This is why RVCs make an excellent design case for the illustration of this chapter's theory. Regarding the below case study, hypothesized data is used for clarification only and may deviate from the actual product or market data.

5.1   Data-Driven Constraint Management of Robot Vacuum Cleaner

Data describe design situations (design concepts and working environments) of constraints. Groups of typical RVC users include young couples, pet owners, and office workers. The differences between these user demographics expose RVCs to various working environments, from apartments and large houses to workshops. Designers also vary in their design philosophies and product differentiation strategies. Variations in customer demographics, working environments, and designer strategies produce a confluence of design situations for RVCs. As introduced in Chap. 3, contextual data helps describe products' working conditions and design features. Two hypothetical scenarios with contextual data are detailed below in Table 1 and

Table 1 Examples of contextual data for different scenarios

Contextual data	Scenario 1	Scenario 2
Timetable	47% 8–10am, 53% 12–2 pm	38% 8–11am, 62% 4–6 pm
Temperature	17–22 °C	20–26 °C
Location	8% kitchens, 67% living rooms, 15% bedrooms, 10% balcony	61% workshops, 29% rest areas, 10% corridors
Passenger	51% adults, 24% cats, 25% dogs	67% designers, 17% managers, 16% operators
Items in space	24% fibre furniture, 31% wood furniture, 8% wool carpet, etc.	26% working benches, 32% manufacturing machines, 12% furniture, etc.
Notification	13% cleaning reminding, 64.1% cleaning statuses, 4.2% software updates, etc.	16% of cleaning reminding, 52% of cleaning statuses, 3.6% software updates, etc.
Interaction with peer products	17% with smart home controllers, 41% with smartphones, 22% with virtual walls, etc.	47% with centre controllers 26% with smartphones, 8% with the security room, etc.
Service output	41% of debris suction, 12% of floor mopping, 24% of data supply, etc.	61% of debris suction, 8% of floor mopping, 13% of data supply, etc.
Resources received	During cleaning, 27% hairs, 8% leaves, 63% dirt, etc.	During cleaning, 12% aluminium scarps, 38% wood powders, 9% steel particles, etc.
Demography	28.6 and 27.2 years old couple, 12.6 h home-staying time per day, 7.9 and 9.1 h sleeping time per day, etc.	71% are between 28 and 45 years old, 31.6% receive bachelor education or above, 79% are male, etc.
Mood	48% normal, 41% joyful, 7% anxious, etc.	67% feel normal, 11% feel cheerful, 8% feel anxious, etc.
Using habits	14% clean once a week, 82% clean at mid-level, 23% mop after vacuum, etc.	72% clean every 4 h, 86% clean at high-level, 86% three or more RVCs work together, etc.
Knowledge of products	6.8 years' experience of smartphones, 2.7 years' experience of smart home controllers, 3.1 years' experience of RVCs, etc.	10.8 years' experience of smartphones on average, 48% have experience in remote control, 3.1 years' experience of RVCs on average, etc.
Energy generation	5–36 V voltage generated, 0.017–0.043m^3/s airflow, 14.63–22.94 W kinematic power, etc.	5-36 V voltage generated, 0.028–0.059m^3/s air flow, 15.72–25.71 W kinematic power, etc.

(continued)

Table 1 (continued)

Contextual data	Scenario 1	Scenario 2
Computation	44% are about cleaning management, 37% are about operation management, 19% are about navigation, etc.	52% are about cleaning management, 22% are about operation management, 10% are about navigations, etc.
Movement	7.6% acceleration, 6.8% braking, 79% linear movement, 13.7% rotation, etc.	5.6% acceleration, 6.2% braking, 85% linear movement, 11% rotation, etc.
Energy consumption	47% are dissipated by heat, 42% are transferred to air and debris, 61% RVCs consume more than 0.63KWh for every cleaning, etc.	53% are dissipated by heat, 47% are transferred to air and debris, 76% RVCs consume more than 1.1KWh for every cleaning, etc.

used to illustrate the RVC case study. Scenario 1 represents young couples with pets who use an RVC for cleaning their apartment, and Scenario 2 represents a team of designers who use RVCs for cleaning their design workshop.

Constraint boundaries exist in different forms. Table 2 highlights some examples of RVC constraints with specified boundaries. Constraints cover a wide range of factors that affect FR achievement, resulting in a wide variation of boundary constraints. Constraint boundaries are often quantitative, such as maximum speed

Table 2 Examples of constraints of robot vacuum cleaners

Number	Design constraints	Boundaries
DC_1	Voltage	36 V
DC_2	Material of Shield	Non-Toxic, insulator
DC_3	Maximum speed	0.68 m/s
DC_4	Maximum weight	3.96 kg
DC_5	Maximum diameter	353 mm
DC_6	Maximum noise	60db
DC_7	Minimum suction power	16 W
DC_8	Minimum battery volume	2600 mAh
DC_9	Minimum signal strength	−85 dbm
DC_{10}	Room size	Roughly 190 m^2
DC_{11}	Furniture dimensions	Floor clearance 95 mm and feet clearance 350 mm
DC_{12}	Debris type	No larger and heavier than leaf
DC_{13}	Home base	\
DC_{14}	Dual-mode virtual wall	\
DC_{15}	Users' mobile phone	Smartphone

and minimum suction power. However, many other boundary constraints are qualitative. For example, the shell exterior of an RVC has constraints such as being non-
toxic and electrically insulated for user protection. Some boundary constraints are
so ambiguous they are difficult to define. For example, the RVC peer product known
as a 'virtual wall', which are used to create invisible barriers RVCs cannot cross
in the home, is also a constraint. They directly impact the navigation and cleaning
functions, varying the spatial constraints for RVC operation. However, finding exact
quantitative or qualitative parameters to represent virtual wall boundaries is difficult,
as they depend upon an infinite variation of spatial situations. This high degree of
variability impedes the precision by which boundaries are defined.

Design situations also impact boundary constraints. Examples of constraints
shown in Table 2 suit RVCs in Scenario 1. When designing RVCs for Scenario
2, some constraints must be modified. For example, as shown by contextual data in
Table 1, debris for Scenario 1 consists of hairs, leaves, and dirt. A minimum suction
power of 16 W and battery volume of 260 0mAh are sufficient for 70 min of cleaning
time. In Scenario 2, the types of debris include wood powders, aluminium scraps,
and steel particles. As the density and amount of debris in Scenario 2 will be greater
than Scenario 1, the minimum suction power and battery volume should be raised to
27 W and 4200 mAh, respectively, to ensure adequate cleaning performance.

Data provides designers with invaluable insights into how constraints influence
FRs. FRs often have a direct effect on constraints. For example, Wi-Fi signal strength
significantly affects the FR 'transfer data to the smartphone'. As shown by Fig. 5, Wi-
Fi strength correlates with data transfer speed, where higher Wi-Fi strength values
result in faster transfer speeds. The Wi-Fi signal strength generally sits between −110
dBm and −30 dBm. When the signal strength is above −85 dBm, the data transfer

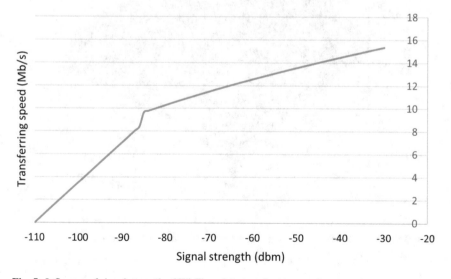

Fig. 5 Influence of signal strength of Wi-Fi on data transferring speed

speed increases parabolically with signal strength. Whereas if the signal strength is below −85 dBm, the Wi-Fi signal becomes discontinuous, dropping rapidly as the signal strength weakens.

In other cases, RVC constraints affect FRs indirectly. For example, a smart home controller may keep sending Wi-Fi signals to control other smart home products but does not directly intercept data transferred from the RVC to the smartphone. However, the signals emitted from the smart home controller may interfere with the Wi-Fi signal strength, which indirectly impacts the transfer speed. Analysis of the data shown in Fig. 5 measures the effect of indirect influences and the extent to which they can be controlled.

It is also possible for multiple constraints to have a compounding effect on FRs. For example, consider that a designer uses photoelectric and laser radar sensors to achieve the FR 'detect debris on the floor'. Under this condition, the colour and size of debris are both constraints that impact the reliability of debris detection, as larger sizes are more visibly identifiable, and lighter colours reflect more red light. As shown in the surface plot in Fig. 6, data collected from simulations, tests, and practical use cases enables designers to plot how different debris colours and lengths impact their detectability. It is also possible to create 4D plots if floor colour is considered a constraint. Such plots help designers examine the impact of constraints on FR achievement.

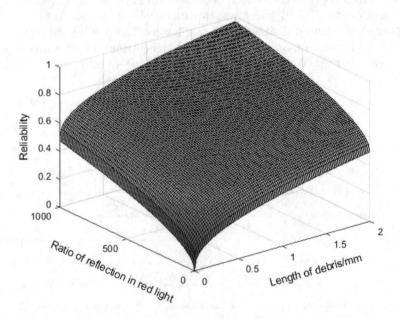

Fig. 6 Influence of debris length and colour reflection ratio on debris detections

5.2 Data-Driven Complexity Management of Robot Vacuum Cleaner

This section explains how data help designers measure complexities, specify CPM targets and create CPM plans based on the Axiomatic Design Theory (Information Axiom). An example where designers develop nozzles that spray detergent and water for an RVC with mopping function is used to illustrate this theory. These nozzles are given two core functions: 'FR_1: Spray 50 mL water on the floor' and 'FR_2: Spray 2 mL detergent every time'. Due to the high viscosity of detergent, it is recommended to be mixed with water to enhance its flowability. Designers plan to install two separate nozzles, one for spraying 3 mL water (DP_1), and the other for spraying 17 mL of foam (DP_2) (the mixture of 2 mL detergent and 15 mL water). As shown in Fig. 7, this concept is decoupled as FR_2 is controlled by DP_2, whilst FR_1 is controlled by both DP_1 and DP_2. Designers intend to control the tolerance of water spray (FR_1) within 3 ml. Proportionally, tolerance of DP_1 should be within 2 ml whilst that of DP_2 should be within 1 ml. Also shown in Fig. 7, the present concept's nozzle spray volume for DP_1 fluctuates between 32 and 36 mL, whilst DP_2 fluctuates between 16 and 19 mL. Based on the Information Axiom, the reliability of DP_1 and DP_2 are 87.5% and 77.8%, respectively. Therefore, the overall reliability of this concept is 68.1%, with more uncertainty contributed by DP_2 than DP_1.

Based on this result, designers consider the reliability of FR_1 to be unacceptable and aim for a value over 90%. Using a similar concept, the nozzles are improved by

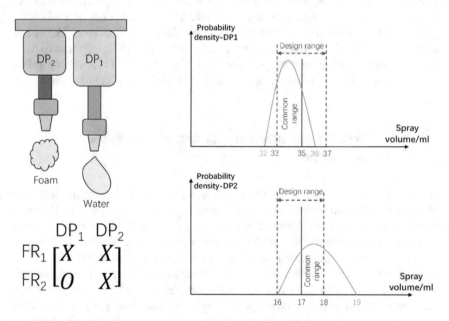

Fig. 7 Complexities of initial plan measured via Information Axiom

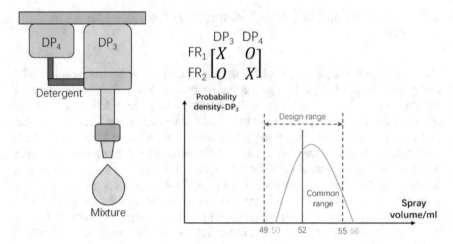

Fig. 8 Complexities of improved plan measured via Information Axiom

tightening the tolerance values. However, the nozzles required to achieve 90% reliability are costly. An alternative solution is concept simplification. Using Axiomatic Design Theory, complexities are reduced by removing couplings among FRs and DPs [1]. As shown in Fig. 8, removing the coupling between FR_1 and DP_2 in the initial concept involves the application of a detergent injector (DP_4) and waterspout (DP_3) for achieving FR_1 and FR_2, respectively. The detergent injector injects 2 mL detergent to the waterspout chamber, which then sprays the mixture of 2 mL detergent and 50 mL water. Here the tolerance of DP_3 must be within 3 mL. The nozzle spray volume for DP_3 fluctuates between 50 and 56 mL, though its tolerance remains similar to that of the initial concept. Based on the Information Axiom, its reliability is calculated as 94.4%; a significant increase in comparison to the initial plan.

5.3 Data-Driven Contradiction Management of Robot Vacuum Cleaner

A data-driven approach to CDM offers designers new ways to identify, evaluate and compare contradictions with different RVC design strategies. A typical RVC contradiction involves the noise generated by airflow at the outlet. This section will use this contradiction example to demonstrate how data support the CDM process. Presently, most RVCs use a combination of a disk and sponge filter to prevent debris from exiting the air outlet. The filters gather debris from the vacuum, which is compressed firmly at the disk filter. However, the compressed debris blocks airflow, generating turbulence and noise. High airflow speeds, caused by greater suction power, result in higher debris accumulation and greater turbulence. Consequently, the noise becomes louder. Therefore, an apparent contradiction is discerned between suction power and noise levels in RVC design.

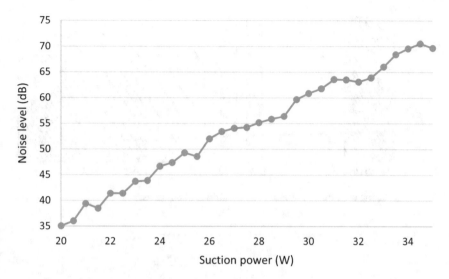

Fig. 9 The contradiction between suction power and noise level

Analysis of the suction power and noise level data enables designers to understand this contradiction. Hypothetically, as shown in Fig. 9, a physical test is conducted for the present concept to collect how the noise level raises with the suction power. In the test, designers increased the suction power from 20 to 35 W, causing noise levels to climb from 35.16 dB to 69.23 dB. Analysis of customer data also reveals that noise levels above 65 dB will dramatically increase users' anxiety levels. From the product and customer data, designers ascertain that an optimal balance between suction power and noise occurs below 33 W operating power. Alternatively, changing filter materials or modifying the outlet will reduce noise generation.

TRIZ helps designers bring up innovative ideas of CDM, and data help designers evaluate the efficacies of their innovative concepts. According to the TRIZ Matrix, this contradiction consists of 'loss of energy' (Feature 29) and 'noise' (Feature 27). The matrix indicates that the contradiction can be resolved by 'Principle 3: Local quality'. The principle can be clarified as 'change an object's structure from uniform to non-uniform, change an external environment (or external influence) from uniform to non-uniform' [21]. As shown in Fig. 10, a candidate concept of this principle is changing single-layer filters into multi-layer filters. Adding disk filter and two sponge filter layers reduce the density compared to the initial concept. When air flows through the disk filters, Disk Filter 2 (which has a lower density) captures large debris particles, and finer debris particles will be stopped by Disk Filter 3. Having multiple filters with an increasing mesh density reduces debris aggregation and the overall airflow turbulence. Sponge Filter 2 slows the airflow speed, which reduces shock at the final filter, which filtrates microorganisms. Once a concept design is complete, designers create a prototype and conduct tests for monitoring its efficacy. Figure 11 presents a quantitative comparison between the initial and improved concepts.

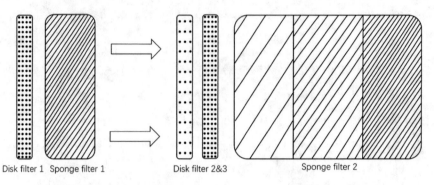

Disk filter 1   Sponge filter 1   Disk filter 2&3   Sponge filter 2

Fig. 10 Changing the structure of filters from uniform to non-uniform

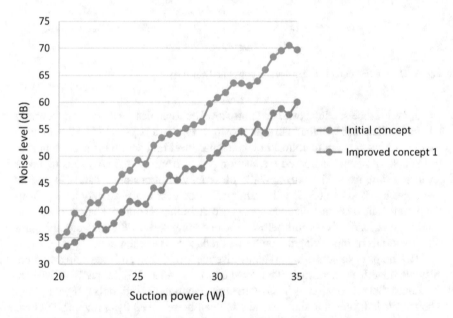

Fig. 11 The comparison between the initial concept and the improved concept

Using information obtained from product and customer data, designers formu-late and compare multiple candidate CDM solutions. The TRIZ Matrix suggests 'Principle 9: Preliminary anti-action' to resolves the contradiction. The principle is defined as 'if it will be necessary to do an action with both harmful and useful effects, this action should be replaced with anti-actions to control harmful units' [21]. A candidate concept of this principle is the installation of sound-absorbing foams at the outlet. Installing foam along the wall outlet acts as a buffer to shocks caused by soundwaves. The improved concept is also tested, with collected data shown in Fig. 12. These results indicate that sound-absorbing foams (Improved Concept 2)

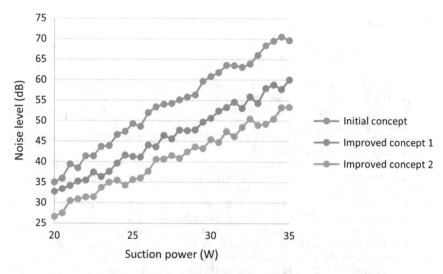

Fig. 12 The comparison among initial and improved concepts

have significantly lower noise levels than the other concepts. Although contradiction resolution was useful in reducing the noise level, other factors such as cost, uncertainties, and physical properties influence designers' decisions.

6   Conclusion

This chapter introduces how data can support CIs. The CI methodologies strive to achieve three critical objectives in design process. First, data fathom constraints, complexities, and contradictions. Second, data delineate the influence of constraints, complexities, and contradictions on FRs. Third, data specify the CI design targets in quantitative forms. For CTM, data are used to describe design situations of constraints, define boundary conditions of constraints, and identify constraint-FR and constraint-constraint couplings. For CPM, data enable improved measurement of design systems, CPM target specification, and the extent of complexities contributed from different sources. For CDM, data uncover contradictions between design features. This assists designers to evaluate various CDM strategies. The CI process is becoming increasingly difficult due to the growing product sophistication and rapid market changes. Therefore, designers must manage the design process in a more analytical, data-driven way or risk obviation in a globally competitive market.

References

1. Suh, N. P. (2001). Axiomatic design: Advances and applications (The Oxford series on advanced manufacturing).
2. Onarheim, B. (2012). Creativity from constraints in engineering design: Lessons learned at Coloplast. *Journal of Engineering Design, 23*(4), 323–336.
3. ElMaraghy, W., ElMaraghy, H., Tomiyama, T., & Monostori, L. (2012). Complexity in engineering design and manufacturing. *CIRP Annals, 61*(2), 793–814.
4. Savransky, S. D. (2000). *Engineering of creativity: Introduction to TRIZ methodology of inventive problem solving.* CRC Press.
5. Cox, J. F., III., & Spencer, M. S. (1997). *The constraint management handbook.* CRC Press.
6. Smith, R. P., & Eppinger, S. D. (1997). A predictive model of sequential iteration in engineering design. *Management Science, 43*(8), 1104–1120.
7. Liu, A., Wang, Y., Teo, I., & Lu, S. (2019). Constraint management for concept ideation in conceptual design. *CIRP Journal of Manufacturing Science and Technology, 24*, 35–48.
8. Hernandez, N. V., Shah, J. J., & Smith, S. M. (2010). Understanding design ideation mechanisms through multilevel aligned empirical studies. *Design Studies, 31*(4), 382–410.
9. Okudan, G. L. E., Ogot, M., & Shirwaiker, R. (2006). An investigation on the effectiveness of design ideation using TRIZ. In *International Design Engineering Technical Conferences and Computers and Information in Engineering Conference* (Vol. 42584, pp. 953–961).
10. Laine, P. M., & Vaara, E. (2007). Struggling over subjectivity: A discursive analysis of strategic development in an engineering group. *Human Relations, 60*(1), 29–58.
11. Lu, S. C., & Liu, A. (2011). Subjectivity and objectivity in design decisions. *CIRP Annals, 60*(1), 161–164.
12. Coyne, R. D. (1991). Objectivity and the design process. *Environment and Planning B: Planning and Design, 18*(3), 361–371.
13. Matta, A. K., Raju, D. R., & Suman, K. N. S. (2015). The integration of CAD/CAM and rapid prototyping in product development: A review. *Materials Today: Proceedings, 2*(4–5), 3438–3445.
14. Tao, F., Qi, Q., Liu, A., & Kusiak, A. (2018). Data-driven smart manufacturing. *Journal of Manufacturing Systems, 48*, 157–169.
15. Pahl, G., & Beitz, W. (2013). *Engineering design: a systematic approach.* Springer Science & Business Media.
16. Liu, A., Lu, S. (2013). Lessons learned from teaching axiomatic design in engineering design courses. In *Proceedings of the Seventh International Confer- ence on Axiomatic Design* (pp. 99–106).
17. Suh, N. P. (2005). Complexity in engineering. *CIRP Annals, 54*(2), 46–63.
18. Frizelle, G., & Woodcock, E. (1995). Measuring complexity as an aid to developing operational strategy. *International Journal of Operations & Production Management.*
19. Van Eijnatten, F., Putnik, G., Sluga, A. (2007). Chaordic systems thinking for novelty in contemporary manufacturing. *CIRP Annals –Manufacturing Technology, 56*(1):447–450.
20. Ilevbare, I. M., Probert, D., & Phaal, R. (2013). A review of TRIZ, and its benefits and challenges in practice. *Technovation, 33*(2–3), 30–37.
21. Al'tshuller, G. S. (1999). *The innovation algorithm: TRIZ, systematic innovation and technical creativity.* Technical innovation center, Inc.

Chapter 5
Blockchain-Based Data-Driven Smart Customization

Abstract With the rapid development of cyber-physical systems (CPS) and big data, product customization is evolving from traditional mass customization to data-driven smart customization. Data-driven smart customization is a new paradigm that highly emphasizes the reconfiguration of products in the use phase. This paradigm calls for a new architecture to enhance collaboration to perform customization. This chapter proposes a blockchain-based data-driven smart customization framework, in which blockchain functions to maintain the evolved customization data and decentralized consensus among stakeholders. A case study of smart vehicle customization is conducted to demonstrate its efficacy.

Keywords Customization · Blockchain · Smart product design

1 Introduction

Faced with increased market competitiveness and more demanding customers, manufacturers must adopt mass customization (MC) to maintain their competitive advantage. MC is a product development strategy that aims to increase product variety without compromising economies of scale [1]. For the past few decades, customization practices have experienced rapid improvements and adoption by many industries, including electronics, automotive, food, and construction. More recently, the intelligent transformation of manufacturing systems has incorporated cutting-edge technologies such as smart products, big data analytics, and machine learning (ML) into product customization processes. Consequently, the design process has become increasingly data rich and technology intensive. For example, manufacturers use smart products to remote monitor their factories, which removed many geographic and spatial constraints of organizing manufacturing operations [2]. The impact of these technologies on how businesses operate is also creating new realms of possibility for product customization, and a step change in how they deliver value to customers. As a result, product customization is experiencing a revolution from conventional MC and the personalization of products en masse, to smart customization, where products are specifically tailored to individual customers' needs [3].

In this new paradigm of smart customization, large volumes of data are used as evidence to determine customer requirements, trigger customization activities, and validate the customization quality. Notably, it is not designers but customers who are the main producers, owners, and consumers of this data. The large volume and complexity of data require new infrastructure for its collection, validation, and management. Furthermore, reliable infrastructure ensures that designers and customers develop mutual trust for collecting and sharing data. Blockchain is among the most promising technologies to support this paradigm shift since it provides immediate and immutable data recording and sharing [4]. Additionally, blockchain verifies customization data in a decentralized manner, which helps build trust between involved stakeholders [5].

This chapter proposes a blockchain-based framework for data-driven smart customization. First, Sect. 2 introduces the primary customization paradigms and their key challenges. Section 3 explains blockchain's functionality and capability, and the suitability of applying blockchain technology. Next, Sect. 4 proposes a blockchain-based data-driven smart customization framework. Its key modules, technical features, and operational processes are discussed in detail. Section 5 demonstrates the application of the proposed framework through a case study on a smart vehicle. Finally, Sect. 6 concludes the main contributions of this chapter and highlights the most vital areas for future research.

2 Overview of Customization

MC is a product development strategy that aims to fulfil individual customers' needs with near-mass production efficiency [1]. The prevailing MC practice increases product variety and controls manufacturing and development costs through modularity and commonality in product design. As MC can effectively increase product variety without compromising economies of scale, this method has achieved significant commercial success in the market. Recent advances in the Internet of things (IoT), information communication technologies (ICT), and information technologies (IT) have steered conventional MC towards a more connected and intelligent paradigm, defined as data-driven smart customization [3, 6]. Compared with the MC paradigm, which solely addresses customer needs during an early phase of the product lifecycle, data-driven smart customization offers designers the flexibility to fulfil customers' evolving needs.

There are three main differences between MC and smart customization, which are summarized in Table 1.

Customer needs: MC usually targets customers' explicit needs, while smart customization emphasizes fulfilling implicit customer needs. Explicit needs are those that customers express clearly, for example, a car customer may state explicitly that he or she needs sport seats. Whereas implicit needs are those which the customer does not yet understand they need or express, for example, a customer may not know that a more comfortable seat design will improve their driving experience, but this need

Table 1 Comparison between mass customization and smart customization

	Mass customization	Smart customization
Customer needs	Explicit customer needs	Implicit customer needs
Configurations	Configuration before use	Reconfiguration in the use phase
Collaboration	Centralized	Decentralized

lies at the root of their discomfort when driving. The prevailing method for eliciting customer needs is through structured questionnaires. Moreover, designers segment customers into groups by identifying their explicit characteristics such as demographic, geographic, and psychological characteristics. The issue with this approach is that it limits what customers' needs are to explicit needs based on pre-defined metrics. Customer needs are often ambiguous and vary based on where they are used and users' preferences. Defining implicit needs has been a significant challenge for designers, as these are often unknown and address the most unique and innovative areas for product improvements. A key advantage of smart customization is its ability to comprehend these implicit needs by analysing individual customer behavioural data. The advances in ICT, IoT have made customers more connected to designers. A variety of data collected by smart products in the use phase provides an opportunity to understand implicit customer needs by analysing user behaviours, user cognitions, and usage context.

Configuration: Smart customization also enables product modules to be reconfigured during their usage phase to fulfil evolving customer needs. Conventionally, customization is realized by configuring different physical modules. The way in which these modules are assembled creates various dependencies and constraints on future alterations. Therefore, after manufacturing, it is very difficult to reconfigure a product's design, since the design of one module requires a redesign of dependent modules. However, the increased popularity of smart products has meant that electrical and software modules play a more dominant role in performing product functions. Electrical and software systems are inherently function specific with little or no physical constraints, which increased product adaptability. Therefore, smart products can realize the reconfiguration post-manufacture through software updates and Internet-based services.

Collaboration: In MC, it is the dominant manufacturer's role to organize and collaborate with other stakeholders to produce a product. However, in smart customization, stakeholders will collaborate in a more self-sovereign and decentralized manner [5]. As customer needs in the use phase are fleeting, relying on dominant manufacturers is becoming time inefficient in capturing those opportunities [5]. In the smart customization paradigm, customers collaborate with locally distributed customization resources to perform reconfiguration tasks. Dominant manufacturers will design open-architecture products and define interface standards to allow potential suppliers to provide individualized modules. Therefore, in the smart customization paradigm, customization activities are not only led by certified large enterprises, but the participation of various SMEs will also significantly contribute to the prosperity of the whole industry.

Despite the many benefits of data-driven smart customization, there are several major hurdles that must be overcome for its widespread adoption. First, data-driven smart customization focuses more on the reconfiguration during the use phase, which requires proactive collaboration between users, designers, and manufacturers. However, it is challenging to build and maintain mutual trust between stakeholders. In conventional MC practices, it is the dominant manufacturers' responsibility to endorse the reliability of other stakeholders. For example, once a supplier fails to meet the quality requirement defined by customers, it is the dominant manufacturer who takes responsibility. Therefore, the selection of stakeholders in conventional MC is a time-consuming process. However, in smart customization, customers will collaborate with various locally distributed manufacturing resources without dominant stakeholders. In this circumstance, clarifying the responsibilities of each stakeholder during the collaborative customization process is an issue.

Second, centralization is one of the primary obstacles to promote smart customization. In the existing product customization paradigm, customized components are only produced by certified suppliers. Certification often incurs significant costs and time for suppliers to pass all requirements. Generally, it is only the largest companies that can afford this. However, with smart customization, customer needs are shifting from a few popular product functions to a vast number of unique functions. The increased variety of functions requires a more diverse supply chain to produce. Hence, more small- and medium-sized enterprises (SME) can participate in product manufacture. Replacing the time-consuming and resource-intensive certification process with a more flexible method is a challenge that must be overcome for these innovative SMEs to participate in the design process.

Third, the data authenticity of the smart customization process is a key concern of the manufacturing industry. Smart customization requires collaboration with locally distributed manufacturing resources in a data-driven manner. Therefore, any fake or disingenuous data sent by malicious parties affects the accuracy and security of the information communicated between stakeholders. For example, manufacturers may create misleading claims about competing products, or individuals access confidential information about product users. Therefore, verifying data authenticity is becoming increasingly vital for the proliferation of smart customization practices.

Blockchain technology addresses the issues associated with smart customization in four primary ways. Firstly, blockchain technology ensures that each stakeholder can receive and share data without a centralized authority [7]. Due to its distributed system architecture, data uploaded on the blockchain can be shared with and accessible to all stakeholders. Secondly, the immutability of blockchain data enhances its traceability, allowing stakeholders to verify data independently [4]. Thirdly, smart contracts, a self-executing application that runs on the blockchain, offer far greater flexibility of financial transactions. For example, one party will automatically be charged based on the extent to which they fulfil a contract's requirements. Finally, blockchain technology guarantees data authenticity in cyberspace, eliminating instances of data forgery.

3 Overview of Blockchain Technology

3.1 Blockchain and Key Capabilities

Blockchain refers to a data structure formed by a list of ordered data or transaction lists, known collectively as blocks, with each single block linked cryptographically by irreversible time stamps. Algorithmic rules and cryptographic techniques, such as digital signatures, are used to authorize transactions. The chain of transactions between these blocks is what forms the 'blockchain'. Through this mechanism data stored in the blockchain cannot be modified.

A typical blockchain system also includes a network of nodes, and a network protocol. Nodes store blocks of data, and they could be any kind of devices, such as computers and servers. Nodes on a blockchain system are connected to enable data on each node to be updated. The network protocol defines the rights of data communication and verification across nodes in the blockchain network. Theoretically, data recorded in the blockchain is tamper-proof. Each block in a blockchain is 'chained' back to the previous block, by containing a 'hash' representation of the previous block. A 'hash' is an algorithm that converts arbitrary sized data to fixed sized data. Thus, historical transactions in the blockchain may not be deleted or altered without invalidating the chain of hashes. Combined with computational constraints and incentive schemes on creating blocks, this prevents tampering and editing of information stored in the blockchain.

Compared with conventional databases, blockchains have several unique capabilities to facilitate smart customization:

Data immutability: Ensuring data immutability is an effective way to build mutual trust among stakeholders. Each block in a blockchain contains a cryptographic hash of the previous block and an irreversible timestamp, and it will become immutable when data is recorded. If a transaction error occurs, a new transaction must be used to reverse the error, with both transactions made visible. As no stakeholder can tamper with data after it has been recorded, data recorded on the blockchain becomes a trusted information source.

Data transparency: All stakeholders on the blockchain network have access to a data record copy, allowing them to verify data integrity without the need for third-party intermediaries. Therefore, incorporating the blockchain into product customization eliminates the need for time-consuming certification processes, as verification processes are decentralized. This enables SMEs to collaborate with manufacturers or even directly with customers.

Decentralized consensus: Through a distributed system architecture and smart contract, blockchain enables consensus-based transactions where agreements on stakeholders are established without requiring a central integration point. Therefore, no privileged stakeholder exists in the blockchain, and all stakeholders on the blockchain must agree that a transaction is valid by verifying related data. The decentralized consensus reduces the cost of verifying product quality and enables SMEs to compete with large companies.

Traceability: The degree of traceability is facilitated by blockchain technology. First, data uploaded on the blockchain are timestamped, enabling all stakeholders to trace its origin. Second, digital signatures have been used in blockchain to sign smart contracts and trigger transactions. This allows the data on the blockchain can be traced back to verified data generators.

3.2 Blockchain Suitability Evaluation Framework for Customization

The cost-effective implementation of blockchain technology in customization systems requires manufacturers to evaluate its suitability for a product. Manufacturers must assess whether using blockchain technology generates more value to customers than the cost of its implementation. A simple five-step evaluation for determining the cost-benefit of blockchain technology compared with a conventional database shown in Fig. 1.

Step (1): Are various stakeholders involved? The value of blockchain is fully exploited when serving numerous stakeholders, especially those who do not trust each other. When customers only need to select final products, such as cosmetic customization, a conventional database is often more cost-effective. However, when customers must interact with many stakeholders, such as manufacturers, governments, suppliers, distributors, retailers, and service providers, which are physically and informationally isolated, blockchain becomes a particularly valuable proposition.

Step (2): Is data synchronization required? Data synchronization occurs when stakeholders have access to consistent and immutable information across all customization activities. For customization projects that require customers to collaborate in the design process, full data synchronization is essential. Data synchronization ensures all stakeholders are informed of customer requirements quickly, avoiding delays or increased production costs. Furthermore, it offers a fast and trusted way for all stakeholders to share or receive vital customization information to ensure aligned quality and progress commitments.

Step (3): Is data confidentiality required? Data confidentiality refers to the level of access to information each stakeholder requires during the customization process. This provides a standard for who has access to what, not simply that all data is accessible by all parties. By ensuring each stakeholder agrees on their required level of access to customization information, greater trust between these stakeholders increases, leading to enhanced collaboration and efficiency.

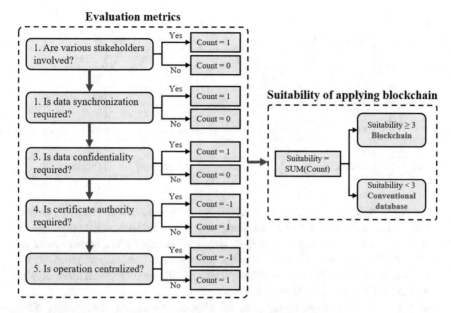

Fig. 1 Blockchain suitability evaluation framework for customization

Step (4): Is certificate authority required? A certificate authority is a monopoly or oligopoly service provider with the accepted credibility to certify other organizations. Many customized products must adhere to legal and specific industrial standards. In some cases, the customization process must occur under the supervision of trusted authorities to ensure compliance with these standards. For such products, the blockchain is generally not recommended. However, for products with open architectures in which many stakeholders contribute to product developments, blockchain technology offers many advantages.

Step (5): Is the operation centralized? Centralized operation occurs when decisions are made by a specific leader or location. Conventional MC is a typical example of centralized operations, where the dominant manufacturer functions manage an entire customization project. Centralized operations provide advantages for risk control, and economies of scale. However, for the customization that needs to operate under high product variety demand, blockchain could be a promising technology to control risks in customization project.

Gilmore and Pine defined four distinctive approaches to MC, namely, transparent, collaborative, adaptive, and cosmetic customization [8]. The proposed framework is used to evaluate blockchain's suitability for those mass customization approaches. Table 2 gives the summarization of the evaluation results based on the five questions from the framework.

Approach 1: Collaborative customization: Collaborative customization involves individual customers to participate in the design process. This approach requires

Table 2 Blockchain suitability to four MC approaches

Metrics	MC approaches			
	Collaborative	Adaptive	Transparent	Cosmetic
1	Yes	Yes	No	No
2	Yes	Yes	No	No
3	Yes	Yes	No	No
4	No	No	No	No
5	No	No	Yes	Yes
Result	Blockchain	Blockchain	Conventional	Conventional

active collaboration between various multidisciplinary stakeholders. Hence, synchronization of data generated by stakeholders, who may need to perform tasks concurrently, is critical for its implementation. Data confidentiality is also important for preventing leakages of sensitive commercial information. The customization process itself does not need to follow a set of compliance regulations, making it more suitable to a decentralized blockchain model.

Approach 2: Adaptive customization: Adaptive customization allows customers to modify products during their use phase. Typically, this involves altering a product's function by changing modules. This approach requires the high participation of customers in modular design. Besides, a larger audience of small- and medium-sized enterprises (SME) are required to produce various individualized modules. Therefore, blockchain technology is suitable for adaptive customization.

Approach 3: Transparent customization: Transparent customization offers products or services without letting customers know explicitly. This customization approach is applicable when customer requirements are predictable, and need one or a few stakeholders to monitor routinely. Therefore, transparent customization only involves a small number of stakeholders, no collaboration is required, and is performed in a centralized manner, which is more suitable for the conventional database.

Approach 4: Cosmetic customization: A cosmetic approach offers standard products but provides customers with the perception that products are customizable. For example, a shampoo company may provide different colours of shampoo without changing any other chemical composition. Generally, cosmetic customization is mass production focused, and therefore is more suitable for conventional databases.

4 Blockchain-Based Smart Customization Framework

The following section proposes a blockchain-based smart customization framework, which consists of three modules as shown in Fig. 2. These are the customization module, data-driver module, and blockchain module. The technical features of each module and their contributions to the whole framework's operation are explained below.

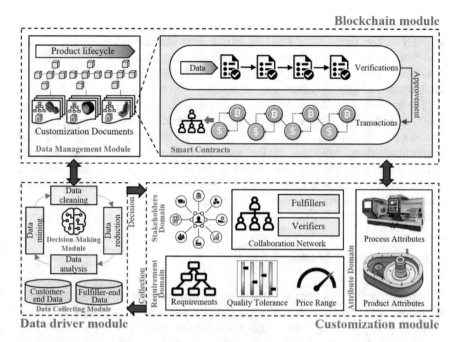

Fig. 2 Blockchain-based smart customization framework

4.1 Customization Module

The customization module accommodates different kinds of customization activities across three domains: stakeholders, requirements, and attributes.

The stakeholder domain accommodates different stakeholders required for the customization project, including service providers, suppliers, distributors, and manufacturers. Each stakeholder possesses different resources to support the customization project. Various stakeholders collaborate with each other to complete specific tasks. When a new customization project is launched, capable stakeholders are matched to form a new collaboration network. Different stakeholders in the collaboration network act as certain requirement fulfillers in a specific part of the project. In meanwhile, all stakeholders also work as verifiers concerning the quality of the task. Each verifier has equal voting rights to determine whether the task meets customer requirements. In this way, distributed customization resources can collaborate with each other in a decentralized manner.

The requirement domain accommodates the customization requirements of a product against quality tolerances and production costs. It involves a hierarchical structure of customization requirements raised by customers. Each requirement is assigned to stakeholders with the necessary capability for its achievement, and assessed based on a determined cost-benefit. A viable quality tolerance is usually agreed upon to meet the minimal degree of requirement fulfilment expected by customers. Of equal import is also understanding the price range which customers will pay for the degree of requirement fulfilment. At the end of a project, a verification process where each stakeholder's requirements are assessed against pre-defined quality and price tolerances occur. If they fail to fulfil these pre-agreed tolerances, the transaction between stakeholders is halted.

The attribute domain accommodates various attributes derived from customized products or customization process. Product attributes involve the functional and non-functional attributes that customers expect a product or service to possess. Whereas the customization process involves the process attributes raised by stakeholders, such as delivery time and cost fluctuations. Such attributes are possible to be measured by sensor devices, and are made verifiable by other stakeholders. The requisite data is then broadcast to explain the extent of its completion.

4.2 Data Driver Module

The data-driver module aims to extracts useful insights from data generated during the customization process. It represents a shift towards more automated or 'smart' customization processes, where data-driven insights identify new possibilities for increased efficiency. The data-driver module separates the function of data collection and decision-making into separate modules, which are explained below.

The data collection module consists of various technologies and techniques for collecting data during the smart customization process. Collected data is classified as either customer-end or fulfiller-end data. Customer-end data relates to the prediction and validation of customer requirements. Typically, customer-end data can be further classified as sensor data, Internet data and user-generated data. Sensor data reflects usage contexts and product performance, which can help to predict customer satisfaction. Internet data are the data collected from websites or other agents connected to the Internet. User-generated data is created by users online, such as online customer reviews, group discussions, and product testings. Fulfiller-end data is generated by resources, processes, and overall quality of customization projects. Resource data pertains to the materials and equipment used in the customization project. Process data are characterized by process variables and logistics variables, including the data related to tooling, setup, routing, production plans, and deliveries.

The decision-making module functions to predict emerging problems, evaluate their potential impact, and recommend solutions. Such problems include new customer requirements, market fluctuations, and resource constraints. The module involves processes that convert raw data into valuable information for decision-making. The process starts data cleaning, where missing values, duplicated data, and errors are removed from the main dataset. Then, a process of data reduction occurs where large volumes of data are converted into an ordered format for easier post-processing. Finally, data analysis and data mining processes extract useful information from the reduced datasets using a variety of machine learning and artificial intelligence techniques. The insights provide quantitative insights which enable more informed and objective decision-making.

4.3 Blockchain Module

The blockchain module applies blockchain technology to support collaboration among distributed stakeholders, enhancing data management and the implementation of the customization process. The blockchain validates, protects, and stores data generated by the data-driver module, guaranteeing its accessibility and reliability for all stakeholders. The data management module stems from product lifecycle management principles, in which data relating to all facets of a product's lifecycle are stored in a time series. The blockchain links each customization project to its prior version, recording details relating to customer requirements, which stakeholders are involved, product attributes, process attributes, and requirement fulfilment levels. Each stakeholder leaves a traceable and verifiable digital signature of their contribution to the blockchain, with each stakeholder given specific permissions to access and upload data.

Furthermore, the blockchain module also facilitates the use of smart contracts (SCs) between stakeholders. SCs are self-executing contracts that verify whether a stakeholder has fulfilled their pre-agreed obligation to the customization process. Without the need for a centralized or governing body, it addresses any payment or execution issues that occur. The SC is embedded with various clauses and rules defining each stakeholder's responsibilities, data verification processes, and transaction conditions. Once a stakeholder uploaded evidence that they have fulfilled their requirements to the SC, the agreed-upon transactions are invoked automatically.

4.4 Systematic Process of Framework

In the proposed framework, the customization project is represented by a sequence of data-driven design tasks. The blockchain functions to verify data, create smart contracts, execute tasks, and finalize transactions. Collaboration between stakeholders is achieved by an iterative four-step process shown in Fig. 3.

First, customers raise a customization requirement. The requirement can be raised by customers explicitly, or be predicted by perceiving the product's conditions

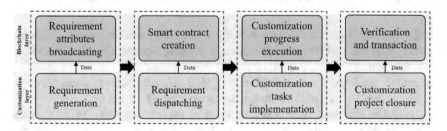

Fig. 3 Customization process framework

(e.g., surrounding environment, running state, user behaviour). The requirement will be decomposed into several product and process attributes. And attributes will be assigned to different stakeholders to complete. The state of attributes is broadcasted on the blockchain. Data are verified by distributed stakeholders to ensure identity and data authenticity. This step aims to initiate a customization project and avoid unfair resource competitions caused by fictitious data.

Second, customer requirements are matched with locally distributed customization stakeholders. Different customization tasks are allocated to different stakeholders based on their location and capabilities. Stakeholders can decide whether to participate in the project based on customer requirements and quotations. Once stakeholders agree to participate in the customization, a SC is created to specify stakeholder's responsibility, quality tolerance, deliverables, price range, and verification methods.

Third, once the customization project is launched, each stakeholder must follow their respective tasks specified in the SC. Several IoT devices are adopted to collect data to reflect the real-time conditions of customization tasks. Based on the real-time data, the data-driver module will evaluate the goal attainment of current progress and predict potential risks (e.g., project delay, unpredicted cost) associated with customization tasks. Then, alternative strategies might be recommended to stakeholders to avoid potential changes. Throughout the project, customization-related data is collected, signed, and broadcasted to the blockchain. This data is made visible to permitted stakeholders, allowing them to verify data authenticity and integrity in a decentralized manner. This approach enables stakeholders to monitor the real-time progress of customization, anticipate potential risks, and adjust objectives to ensure the on-time completion of the customization project.

Fourth, at the end of a customization project, the SC will verify the extent to which each stakeholder's requirement was fulfilled. The verification process is performed progressively and iteratively by a small group of stakeholders in a decentralized manner. Once all stakeholders meet all criteria of the contract, all following transactions are automatically carried out. Also, the transaction record will be broadcast to the blockchain as a new product version.

5   Applications of Blockchain in Smart Customization

Integrating blockchain technology into the smart customization process facilitates greater collaboration between stakeholders along the entire product life cycle. Moreover, the customization operations are made more transparent to customers, fostering more trust and loyalty with manufacturers. The following section discusses three promising applications of the proposed framework above.

Agreement clarification: Blockchain mitigates the contract-breaking risks caused by perception gaps in the conventional customization approaches. Many tasks are regarded as 'black box' by customers in mass customization in terms of inputs, outputs, mechanisms, and controls. As customer requirements are often ambiguous, failing to clarify quality metrics and customer expectations can lead to unperformed obligations. Blockchain helps eliminate such problems at the early design stage of customization. By using SC, customization requirements are quantified and decomposed into a set of metrics and are allocated to relevant product/process attributes. Through an iterative negotiation process, all stakeholders will have a clear agreement of customization requirements, quality tolerance, and price range. During the customization process, various sensors and instruments are adopted to monitor, collect, and upload verification data onto the blockchain platform to allow other stakeholders to verify the quality in a decentralized manner. By monitoring all facets of supplier products, potential quality defects are identified in advance, enabling quality-related risks to be detected, diagnosed, and addressed promptly.

Flexible pricing: In contrast to a conventional fixed-price approach, where a price is agreed upon before a transaction takes place, blockchain-based smart customization enables more flexible pricing arrangements between stakeholders. Here payments are subject to a pre-defined price tolerance, rather than a set price, based on how well they meet or exceed customer requirements. During customization projects, various IoT devices are adopted to collect customization data and broadcast them on the blockchain. Data analytics are used to predict to what extent customer requirements are being fulfilled. At the end of the customization project, by specifying the level of requirement fulfilment, data analytics can calculate the precise price of customization. This mechanism offers stakeholders who put more effort into customization higher rewards and provides greater flexibility to counterbalance resource constraints with customer expectations.

Data sharing service: Smart customization requires large data quantities from heterogeneous sources, which must be shared with a complex network of stakeholders. Ensuring sufficient accessibility of data for all stakeholders was a significant challenge, as conventionally there needs to be a centralized organization and platform where it is stored, processed, and communicated efficiently. Blockchain technology offers two notable benefits for data sharing. First, the decentralization of data sharing removes the need for a governing body and provides all stakeholders with equal access to relevant data. Second, it provides a trusted method for traceability, guaranteeing authenticity and integrity of shared data. The data sharing benefits that blockchain technology offers facilitate far more efficient and reliable processes needed for smart customization.

6 Case Study: Customization of Smart Vehicle's Merging Module

A revolutionary transformation in the automotive industry towards more intelligent and connected vehicles, known as smart vehicles (SVs), is taking place. As SVs are connected to Internet-based communication networks, they can reconfigure their software modules remotely to alter their overall functionality. For example, SV's software can be updated to enhance suspension dampening or reduce energy consumption. As software modules are highly customizable and adaptable, it is foreseen that SV industry may face a high demand for software modules. Against this background, a case study of SV customization is conducted to demonstrate how the proposed framework enhances collaboration between stakeholders and improves the smart customization process. The smart customization of SV's merging module is shown in Fig. 4.

Firstly, customer requirements are identified by analysing user driving behaviours. Based on the statistical analysis of driving history, the SV can predict additional functions that may improve users' driving experience. For example, if an SV detects that a user has issues merging into traffic at a highway entrance, it is possible for a specific module to be developed to address this issue. Requirements of such a module could include the following:

- Detecting traffic on the main road.
- Predicting where the vehicle needs to travel.
- Predicting merging locations.
- Transmit merging intentions to other vehicles.
- Display merging guidance.

With the help of design experts, customer requirements are decomposed into product attributes and service attributes. Product attributes are relevant to the physical embodiment of the SV, such as the components required for the highway merging module. Table 3 presents examples of how product attributes relate to each model. For example, the communication system's function is to exchange information, such as an SV's position, path history, and kinematics, with other vehicles on the main road. The GPS device determines an SV's geographical location, while the onboard radar and

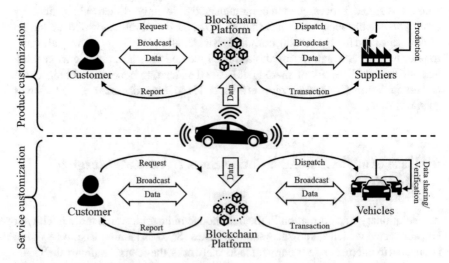

Fig. 4 Highway merging module customization of SV

Table 3 Example of product attributes relating to highway merging module

Product components	Attributes	Reqt
Communication system	System delay	≤ 50 ms
	Protocol	LTE-V2X
GPS device	Location precision	≤ 5 m
Onboard radar	Max distance	≥ 100 m
Wide forward camera	Max distance	≥ 60 m
	Field of view	$\geq 140°$
Narrow forward camera	Max distance	≥ 200 m
	Field of view	$\geq 40°$
Main forward camera	Max distance	≥ 150 m
	Field of view	$\geq 50°$
Processing unit	Computing power	≥ 200 tops
Human-machine interface	Response time	≤ 110 ms

cameras detect nearby vehicles and objects. Another function of the onboard radar is predicting an SV's location when the GPS signal is limited. Furthermore, processing units analyse communication system data to predict upcoming merging locations and provide drivers with instructional guidance. Finally, the human-machine interface projects guidance information to the head-up display.

Service attributes represent those that are controlled outside the physical SV, such as updates on road or traffic conditions generated by entities outside the SV. Table 4 summarizes the main service attributes related to SVs. Road data communication involves providing SVs with accurate road conditions.

The highway merging service requires distributed vehicles to collaborate with SV proactively. Those service attributes are provided in Table 4. Road data communication functions to provide SV with updates of main road conditions. The vehicle give-way service provides the SV with merging assistance by scheduling vehicles on the main roads. The merge position prediction calculates the best time to merge to the

Table 4 Example of service attributes relating to highway merging module

Service components	Attributes
Road data communication	Time of vehicle dispatching
	Availability of road data
	Data granularity and accuracy
	Cost
	Transmission efficiency
Vehicle give way	Response time
Merging position prediction	Prediction accuracy
Merging guidance	Guidance accuracy
	Guidance responsiveness

main road by analysing real-time road data. And merge guidance functions to predict the road path and speed SV should follow to merge in the main road successfully.

Figure 5 illustrates the verification transaction process during the product customization. Initially, the customer raises a requirement for the highway merging assistant module. The requirement is assigned to the relevant specific product components and the suppliers who manufacture them. Each supplier offers a quotation based on customer requirements for product components. Once customer and suppliers reach an agreement regarding product attributes and price, the customization project is launched. In the customization project, each supplier broadcast attribute data to the blockchain, the smart contract will verify the level of requirement fulfilment. Once requirements are fulfilled, the smart contract will initiate payment to each supplier.

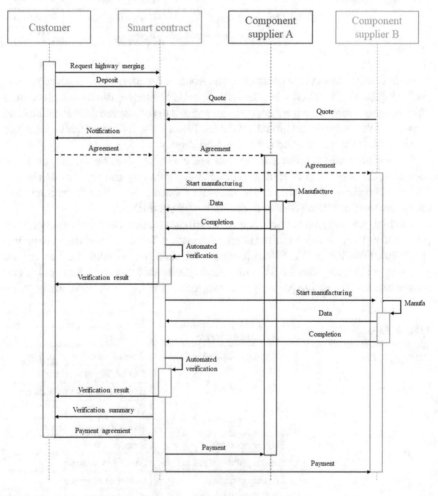

Fig. 5 The verification transaction process of product customization

Similarly, the verification transaction process of highway merging service is illustrated in Fig. 6. In the beginning, the customer requests a highway merging service. The request is dispatched to surrounding vehicles on the main road. The vehicles that are willing to share data and participate are identified. Participating vehicles are classified into two groups. The data-sharing vehicles use equipped sensors, radars, and cameras to search for suitable locations for merging and report information to customers. Their service tasks can be verified by measuring the volume of data reported to customers. The give-way vehicles are responsible for leaving enough gap to allow the customer to merge. And their tasks can be verified by measuring the latitude and longitude control of vehicles. Once the customer is successfully merged in the highway, the smart contract will initiate payment to each vehicle.

Introducing blockchain to the highway merging product-service can benefit many distributed stakeholders. Firstly, blockchain ensures data authenticity on the road. By allowing data to be verified by other vehicles in a decentralized manner, the

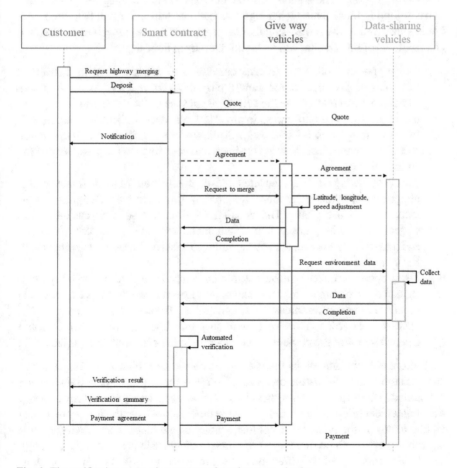

Fig. 6 The verification transaction process of service customization

blockchain can prevent the chance to broadcast false information to seek personal benefits. Secondly, blockchain stimulates vehicle-to-vehicle collaboration through an incentive mechanism. Stakeholders who actively participate in the project are rewarded accordingly. Thirdly, blockchain makes previously confidential data transparent and traceable to the whole network, which improves the quality of product and service that to be offered.

7 Conclusion

The recent years witness the booming emergence of data, which has revolutionized product customization towards a more intelligent and data-driven manner. However, big data cannot serve as eligible contributors unless alleviating the poor collaboration among various stakeholders, insufficient data visibility, integrity, and authenticity. Adopting blockchain technology to smart customization will help the stakeholders involved in the project collaborate progressively and proactively. The main contributions of this chapter can be highlighted from three aspects:

1. The proposed framework enables customers to collaborate with distributed resources at any time to complete a customization project. As no dominant stakeholder can manipulate the collective consensus, the consensus involved in the customization is built and maintained in a fully decentralized manner. Benefit from this, the proposed framework will allow a large number of SMEs to actively participate in the customization business and contribute to the prosperity of the entire industry.
2. The proposed framework ensures the customization of data transparency, integrity, and authenticity along the product life cycle from design to manufacturing and use phase. The detailed requirements of each customization project, along with participating stakeholders, level of requirement fulfilment, and transactions, are executed by the smart contract and stored in the blockchain chronologically.
3. The proposed framework makes customization projects flexible to quality tolerance and price range. This mechanism provides stakeholders more freedom to schedule time and resources under constraints. Besides, the mechanism stimulates the rewards to more successful customization endeavours, which could incentivize more stakeholders to collaborate in the customization project.

There are many limitations that should be considered. Firstly, existing Internet and communication infrastructures cannot support the real-time data transmission mentioned in the article. Besides, there lacks a widely applicable mechanism to verify stakeholder identities on the blockchain. A verifiable digital identity is a key technical feature to ensure the trusted data communication on the blockchain. Therefore, this paper only serves as a preliminary exploration of blockchain in smart customization. The future work lies in two directions. First, more data analysis methods, such as data mining, machine learning, and artificial intelligence, should be developed and incorporated with the proposed framework. The analysis methods that are compatible with multimodal data from heterogeneous devices should be developed. Second, the blockchain can integrate self-sovereign identity (SSI) to enable stakeholders to manage and control their own data. Besides, SSI technology will facilitate privacy when sharing data.

References

1. Tseng, M. M., & Jiao, J. (2001). Mass customisation. *Handbook of Industrial Engineering, 3*, 684–709.
2. Tomiyama, T., Lutters, E., Stark, R., & Abramovici, M. (2019). Development capabilities for smart products. *CIRP Annals, 68*(2), 727–750.
3. Wang, X., Wang, Y., Tao, F., & Liu, A. (2021). New paradigm of data-driven smart customisation through digital twin. *Journal of Manufacturing Systems, 58*, 270–280.
4. Zhang, Y., Xu, X., Liu, A., Lu, Q., Xu, L., & Tao, F. (2019). Blockchain-based trust mechanism for IoT-based smart manufacturing system. *IEEE Transactions on Computational Social Systems, 6*(6), 1386–1394.
5. Liu, A., Zhang, D., Wang, X., & Xu, X. (2021). Blockchain-based customisation towards decentralised consensus on product requirement, quality, and price. *Manufacturing Letters, 27*, 18–25.
6. Zhang, C., Chen, D., Tao, F., & Liu, A. (2019). Data driven smart customisation. *Procedia CIRP, 81*, 564–569.
7. Xu, X., Weber, I., & Staples, M. (2019). *Architecture for blockchain applications* (pp. 1–307). Cham: Springer.
8. Gilmore, J. H., & Pine, B. J. (1997). The four faces of mass customisation. *Harvard Business Review, 75*(1), 91–102.

Chapter 6
Data-Driven Design of Smart Product

Abstract This chapter presents a structured framework that can guide designers to develop smart products with context-smartness, network-smartness, and service-smartness, the definitions of which are explained from the perspective of product development. The framework is developed based on the theoretical foundation of the Situated Function-Behavior-Structure (FBS) ontology. A systematic step-by-step design process is prescribed to leverage various data operations to support design operations involved in the process of designing a smart product with stronger context-smartness, network-smartness, and service-smartness. Some existing design methods and emerging design tools are also incorporated into the framework.

Keywords Smart product · Design methodology · FBS ontology · Context-smartness · Network-smartness · Service-smartness

1 Introduction to Smart Product Design

The rapid development of information, communication, sensing, robotic, and mobile technologies paves the way for the ubiquitous emergence of smart products in the modern society [1]. Typical examples of smart products include smartphone, smart home, smart car, smart appliances (e.g., smart refrigerator, air conditioner, robotic vacuum cleaner, etc.), smart medical device, smart wearable device, and so forth. Majority of smart products are developed either by adapting an existing product in order to make it more intelligent/autonomous (e.g., the transformation of a non-smart refrigerator into a smart refrigerator) or by integrating the functions of multiple products into one more capable product (e.g., the integration of mobile phone, music player, and camera towards a smartphone). Compared to those non-smart products, the smartness of a smart product hinges on its ability to collect, process, transmit, store, analyse, and integrate data.

Smart products can lead to new values at multiple levels. At the product level, a product's smartness can improve the product functionality, performance, and product-user interaction, hence enhancing the customer experience. In particular, unlike those non-smart products that can only submissively respond to interactions initiated by customers, smart products can proactively initiate new interactions and occasionally even drive the interaction with target customers. For example, a smartphone can actively remind the customer about an incoming event, guide the customer to complete an activity, and even perform a task on behalf of the customer. At the industry level, new data obtained by smart products can enable a manufacturer to develop in-depth understandings of target customers, market competition, and product lifecycle towards more informed design decision-making [2]. At the ecosystem level, as more and more products are transformed towards smart products, they can constitute a more connected and self-reinforcing Internet of Things (IoT) through which the network can naturally foster new collaborations across product, stakeholder, and even organizational boundaries. Some smart products and devices can be further interconnected to form even more complex smart systems such as smart factory, smart city, smart warehouse, smart airport, smart farm, smart school, smart hospital, etc.

Product-smartness is a multi-faceted notion. This book focuses on three facets of product-smartness, namely, *context-smartness*, *network-smartness*, and *service-smartness*. Context-smartness refers to a product's quality of intelligence to base on relevant data to interpret a surrounding context, constantly adapt its course of actions within a context, and customize its behaviours against different contexts. Network-smartness refers to a product's quality of intelligence to recognize peer products with complementary functions and behaviours, connect with peer products to exchange information and share resources, as well as collaborate with peer products to perform collective tasks (e.g., model training through machine learning) that cannot be fulfilled by any individual smart product. Service-smartness refers to a product's quality of intelligence to request various value-adding services such as warranty, installation, recharging, maintenance, repairing, return, upgrade, diagnosis, cleaning, recycling, and so forth. On top of those smart products designed for the consumer market, it should be noted that the three kinds of smartness are equally applicable to a variety of autonomous devices (e.g., self-driving car, autonomous drone, and industrial robot) that are deployed in the industrial scenarios (e.g., mining, smart city, and agriculture).

The iterative design of smart products can benefit from artificial intelligence and data science in multiple ways. Firstly, since smart products are to be connected into a network, they naturally require data operations such as data exchange, transmission, and storage. Secondly, through data analytics, it is possible to identify anomalies or outliers with respect to product behaviour, performance, and structure. Thirdly, through machine learning, a smart product can be made more autonomous and capable of predicting. Lastly, customers of certain smart products (e.g., smartphone, smart speaker, and self-driving car) often have higher requirements concerning data security and data privacy. Emerging approaches such as federated learning (i.e., a distributed paradigm of machine learning that is designed to preserve privacy) and blockchain (i.e., a distributed architecture of data management and record documentation) can be leveraged to enhance the privacy and security of smart products.

The remainder of this chapter is organized as follows. Section 2 characterizes a smart product in terms of its context-smartness, network-smartness, and service-smartness. Section 3 presents a theoretical framework of data-driven design of smart products with respect to its key operations, theoretical foundation, and systematic process. Section 4 concludes the chapter and outlines some future work.

2 Characteristics of Smart Product

Based on the definition in the CIRP Dictionary of Production Engineering [3], a smart product can be interpreted as a unique hybrid that integrates cyber-physical system (CPS) with product-service system (PSS). The rise of smart products can be attributed largely to the advances of mobile and sensing technologies. From a technological perspective, smart products are typically characterized by a selective and sophisticated integration of sensors, actuators, processor, software, control technologies, communication technologies, artificial intelligence, etc. Smart products are therefore known for the high degrees of personalization, reconfigurability, autonomy, adaptability, intelligence, and product-service integration [4]. The research on smart products can be approached from different angles such as marketing, management, cognitive science, computer science, industrial design, engineering design, Internet of Things, product development, and so forth.

Indifferent from other artefacts, smart products are essentially designed to satisfy human needs. As shown in Fig. 1, as the intelligence level of smart products continue to increase, it is interesting to explore whether, in what ways, and to what extent a smart product may eventually develop certain 'product needs' that are conceptually analogous to human needs (i.e., different levels of needs included in the Maslow's Hierarchy of Human Needs). For example, would a smart home 'need'

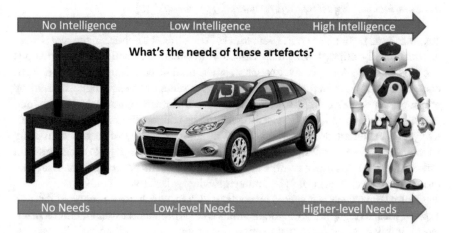

Fig. 1 Evolvement of product 'needs' and 'wants'

a cleaning service, would a robot 'desire' to 'retire' early, would a smart car 'want' to 'communicate' more or less frequently with other cars, and so forth.

In comparison with those non-smart products, a distinguishing feature of smart products is their abilities to communicate, connect, and collaborate with other smart products or service providers within a shared network. As a result, smart products are capable of delivering functions and customizing behaviours that demand sophisticated interactions between a product with customer, environment, and other peer products. Smart product is a multi-faceted notion, and in this chapter, the spotlight is projected to three facets of product-smartness that are closely related to data-driven design, namely, *context-smartness*, *network-smartness*, and service-smartness. Furthermore, as a product's smartness continues to enhance, coupled with the increasing degree of product personalization, it is interesting to investigate whether and to what extent a product will eventually develop its own character (i.e., a set of unique cognitive or behavioural qualities that differentiate a smart product from other smart products with identical technical specifications).

2.1 Context-Smartness of Smart Product

In product development, the notion of 'context' can be interpreted as a set of circumstance factors that instantiate a particular setting, in which the behaviours of a product can be explained, justified, and even predicted. A product context can therefore be described by a selection of variables whose values can portray the context. Accordingly, a smart product is expected to function, behave, and configure differently in various contexts. In other words, against different contexts, a context-smart product should actively adjust its interaction with target customer, peer product, and service provider. In this book, context-smartness is therefore defined as a '*product's ability to perceive, interpret, learn, and integrate contextual information to guide decision-making, adjust behaviors, and adapt structure*' [5].

Generally speaking, context-smartness can be classified into proactive context-smartness and responsive context-smartness. The difference between the two classes is conceptually aligned with the distinction between automatic and autonomous decision-making. A responsively context-smart product can recognize a pre-defined context, respond to interactions initiated by customers or peer products based on preprogrammed rules, inform customers about unknown contextual changes, and adjust product actions purely based on customer commands. In contrast, a proactively context-smart product can actively monitor contextual changes, constantly adapt its behaviours, continuously learn unknown contexts, and make decisions with little or without customer commands. For modern smart products, it is clear that proactive context-smartness is becoming growingly desirable.

Context-smartness is powered by contextual data, which can be acquired through explicit means (e.g., direct communication among product, customer, and environment), implicit means (e.g., customer survey, product review, ethnographic observation, and usage report), as well as statistical means (e.g., data analytics to discover meaningful patterns shared by many products). Contextual data acquired at different time points can be integrated to build a holistic context-smartness. Historical data is useful for context modelling and mining, whereas real-time data is useful for context matching and learning.

Biologically inspired design (BID) is an applicable approach for designing context-smart products [5]. Most biological systems are equipped with extraordinary context-smartness, which is a must-have ability for the sake of Natural Selection. For example, the strong awareness of animals to various risk factors in the wild can be leveraged to design more capable autonomous vehicles. Moreover, some biological systems can maintain a high context-smartness against extreme constraints, which is inspiring for the design scenarios that are characterized by rigorous boundary conditions. Finally, some biological systems can achieve collaborative context-smartness through mutually beneficial symbiotic relationships, which can be leveraged to construct a collaborative ecosystem of smart products.

2.2   Network-Smartness of Smart Product

A network refers to a group of mutually compatible objects (e.g., computers and mobile devices) that are connected to exchange information and share resources based on common communication protocols. Internet (i.e., the network of computers) constitutes one of the most well-known artificial networks in the modern society. Social network (i.e., the network of social actors such as individual people or corporate organizations) represents another massive network. The Internet of Things refers to a network of interconnected physical things including smart products. For example, smartphones, autonomous vehicles, shared bicycles, drones, and smart traffic lights can form a network of smart transportation (i.e., a particular kind of IoT network).

In the digital age, various artificial networks are increasingly merged towards a larger network of networks (NoN), i.e., a particular kind of System of Systems (SoS). For example, the Internet and Internet of Things can be merged through enabling technologies such as cyber-physical system (CPS) and digital twin (DT). Such a network merging introduces both opportunities and challenges to the design of smart products. On the one hand, it enables a smart product to acquire necessary data about surrounding context and relevant services from more diversified channels. For example, a smart air conditioner can query the weather information directly from the Internet through a smart speaker in lieu of self-carried sensors. By doing so, not only the so queried information is more accurate and updated, but also some duplicate resources (e.g., sensors) may be preserved. On the other hand, the network merging will escalate the complexity of protecting data privacy and data security.

Network-smartness refers to the intelligence of a smart product to proactively establish a new network or join an existing network in order to exchange information and share resources with other smart products in the same network. Similar to context-smartness, design inspirations can be gained from biological systems to develop more network-smart products [6]. Some biological systems are featured with strong collaborative behaviours that cannot be otherwise afforded by any individual system. For example, the network-smartness of bees and ants can be investigated to inspire the design of autonomous drones and driverless cars, respectively. Identical to the context-smartness, many biological systems can achieve network-smartness against harsh constraints in the nature. In fact, certain network-smartness is exactly the consequence, or proactive choice, of coping with the Natural Selection. As such, gaining inspirations from biological systems for the sake of design may also improve the competitiveness of smart products against the market competition (that is, conceptually analogous to the Natural Selection).

2.3 Service-Smartness of Smart Product

Smart product involves a number of value-adding services such as maintenance, repair, upgrade, recycling, sharing, etc. Traditionally, the provision of services is primarily intended to extend a product's lifecycle. For example, the service of maintenance can restore a product's functionality, performance, and reliability; the service of recycling can enhance a product's reusability and sustainability; and the service of upgrade can improve a product's compatibility with new software. In the digital age, services are playing more important roles in enhancing product-customer interaction and even affording new functions. Traditionally, the value-adding services are either prescribed by manufacturers based on historical experience or requested by customers as per needed. A service-smart product should however be equipped with the advanced capability of proactively requesting for necessary services. Data science and machine learning have great potentials to reshape how and in what ways services can be integrated into a smart product. For instance, the paradigm of maintenance is evolving from preventive maintenance to predictive maintenance and beyond [7], which is more data driven, condition triggered, and logic regulated.

Service-smartness can be defined as the intelligence of a smart product to proactively identify, request, and choose the most relevant value-adding services with little involvement of human operators. The effectiveness of a product's service-smartness, more or less, depends on its context-smartness and network-smartness. The context-smartness enables a smart product to recognize necessary services for a particular context, detect available service providers in a surrounding environment, and choose the best service against given contextual conditions. For example, a self-driving electric car can identify nearby charging stations for value-adding services such as recharging, repairing, and cleaning. On the other hand, the network-smartness enables a smart product to obtain services from peer products with complementary capabilities within the same network. For example, a network of self-driving cars can communicate with each other to avoid traffic congestions and potential collisions.

Service-smartness plays an important role in promoting product-service integration towards smart product-service systems (SPSS) [8]. Unlike a physical product whose ownership is largely exclusive, services are by nature more inclusive and hence more suitable for sharing [9]. Against the sweeping trend of the Sharing Economy, more and more functionalities can be fulfilled by intangible services in lieu of tangible products. Some examples include shared transportation (e.g., shared cars and bicycles) as well as shared space (e.g., shared housing and office). The service-smartness enables a product to provide functionalities to other customers or peer products in the format of services. In this way, a smart product can be made more sustainable. Not only the product lifecycle can be further extended by proactively requesting necessary services (e.g., maintenance and repair), but also the context-smartness enables a product to identify and possibly rectify unsustainable product/customer behaviours.

3 Data-Driven Design of Smart Product

3.1 Theoretical Foundation of FBS Ontology

A systematic process of data-driven smart product design is proposed. The process is developed based on the theoretical foundation of the Function-Behavior-Structure (FBS) ontology [10], which models engineering design as iterative interactions (i.e., transformation and comparison) among different design entities. Based on the FBS ontology, the essential design entities include requirement, function, behaviour (i.e., expected behaviour and actual behaviour), structure, and description. The relevant design operations that link these design entities include formulation, reformulation, synthesis, analysis, evaluation, reevaluation, and documentation [10]. The FBS ontology can be situated in different contexts to reflect more complex interactions among design entities across different dimensions and in different spaces.

The situated-FBS ontology can navigate designers to cognitively 'travel' across three interrelated worlds, i.e., the expected, interpreted, and external worlds. The expected world is constructed in alignment with a designer's expectation of the functions, behaviours, and structures of a smart product as well as their mutual interactions. Such expectations are framed by designers based on their interpretation, classification, and prioritization of customer voices. Conventionally, the expected world exists largely, if not solely, within a designer's mind, representing a smart product's ideal state. Digitalization makes it possible to build and present the expected world through various digital means.

The external world represents the physical world, where a smart product actually interacts with target customers, external environment, and peer products. Traditionally, designers rely on observation, experiment, and empirical knowledge to understand a product's external world. The emerging technologies such as the Internet of Things enable designers to understand the external world in finer resolution, in greater scale, and even in real time if so necessary.

The interpreted world represents a designer's understandings of how and in what ways a smart product should function, behave, and be structured against potential uncertainty, affordance, constraint, and competition from the physical world. The interpreted world exists in between the expected world and the external world. It is constructed based on two-way interaction between designers with various objects in the external and expected worlds. On the one hand, in the interest of promoting innovations, as much as possible, a designer should depend on design ideality (formed in the expected world) to drive the direction of decision-making. On the other hand, a designer should equally acknowledge the limitation, working principles, and physical laws that are extracted from the external world. It is interesting to note that, due to the increasing popularity of digital twins, more and more smart products nowadays also exist in the digital world—a digital reflection of the physical world. In some sense, the digital world can be regarded as a hybrid between the external world and the interpreted world [11, 12].

Three kinds of inspirations that can be extracted from data to support a designer's decision-making in regard to different FBS operations.

- Type (I): Within the expected world, data about target customers, market competition, and relevant social realities (e.g., social trend, cultural difference, a product's lifestyle meaning) can inform designers to adjust their expectations of what and to what extent a smart product can or cannot achieve. This type of data inspiration plays important roles in the task clarification phase and conceptual design phase of an engineering design process (i.e., product planning in product development [13]).
- Type (II): Within the interpreted world, data about design conceptualization, prototyping, and testing can inspire designers to make necessary design iteration, refinement, and optimization. From the perspective of data science, this kind of data inspiration involves establishing back-and-forth mappings between a new product and existing products with respect to their shared functions, constraints, behaviours, structures, working principles, etc. This kind of data inspiration is most useful in the detailed design and testing phases of an engineering design process.
- Type (III): Within the external world, data about a smart product's physical states and the surrounding environment can inform designers to reconfigure the product structure, customize the product behaviours, and prescribe value-adding services. This kind of data inspiration is most useful in the usage and service stages of a product's lifecycle and the production ramp-up phase of an engineering design process.

3.2   Process of Data-Driven Smart Product Design

As illustrated in Fig. 2, a systematic process of data-driven smart product design includes multiple steps. This process comprises a set of key design operations such as functional formulation, constraint specification, behaviour analysis, structure synthesis, structure coevolution, structure improvement, design documentation, etc. A set of data operations from data science and machine learning are incorporated into different steps of the proposed design process.

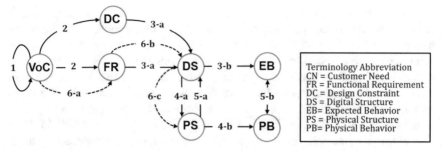

Fig. 2 Systematic process of data-driven smart product design

Step (1): Extract information from customer voices

Smart product is designed to satisfy customer needs. The design process therefore begins with understanding and extracting information from customer voices. Customer voice is by nature a multi-faceted notion that includes customer complaint, confusion, surprise, problem, need, want, preference, emotion, journey, expectation, reflection, perceived quality/risk, etc. Different facets of customer voice can be investigated to support different design operations. For example, information about customer preference, complaint, and perceived quality are useful for the design operations of concept evaluation, service design, and quality assurance, respectively.

For example, customer voices concerning various smart appliances (e.g., robotic vacuum cleaner) can be readily collected from the Internet, where a large volume of user-generated content (UGC) exists in various e-commerce platforms (e.g., Amazon), product forums, social media (e.g., Twitter), and online video platforms (e.g., YouTube). Different from customer voices obtained through traditional design methods, customer voices directly obtained from the Internet are largely unstructured data [14]. As a result, such dataset often contains inconsistent, if not contradictory, information. It is therefore necessary to perform a series of data operations (e.g., data cleaning, data integration, and data mining) in order to extract relevant information [15]. For example, sentiment analysis can be performed to comprehend customer satisfaction with different product features [16], and recommendation systems can be employed to recommend new products to target customers and recommend new functions to a target product [17].

Customer voices may vary significantly in different contexts. Take the design of a robotic vacuum cleaner (VRC), for example, the function of 'cliff detection' is generally regarded important in the US market, where VRC is typically deployed in a house environment that has stairs. On the other hand, the same function becomes less important in other markets where the VRCs are used in various apartment environments without stairs. In another example, customers who own pets tend to have unique customer needs (i.e., cleaning pet hair) and expectations. The customer satisfaction with VRC is also affected by the quality of services, including not only regular services (e.g., warranty, upgrade, and maintenance) provided by the manufacturer but also services (e.g., packaging and delivery) provided by the retailer or distributor.

Among different facets of customer voices, designers are most interested in customer need and customer want. Traditionally, customer needs are solicited by means of survey, individual interview, focus group interview, contextual inquiry, ethnographic observation, etc. These design methods more or less depend on the expertise of designers to collect, aggregate, and extrapolate customer voices manually. Data science makes it possible to directly collect and analyse massive data about customer need/want based on the Internet and the Mobile Internet with respect to a customer's transactional record, searching history, user-generated content, location, journey, online behaviours, etc. Moreover, a variety of new approaches can be employed to analyse customer voices autonomously by machines.

In summary, the input of this step is the raw data of customer voices, whereas the output is a selection of customer needs. For the design of smart products, it is especially important to understand the diversified customer needs against different contexts, the relevant customer needs concerning peer products connected into the same network, and the customer needs of product-related services.

Step (2): Formulate functional requirement and design constraint

Next, information extracted from customer voices is mapped to functional requirements and design constraints. Such a mapping process is typically regarded as a 'translation' process through which the vague customer language is translated into specific engineering language. Relevant design methods for functional formulation include Kano Model [18], Quality Function Deployment [19], Design Structure Matrix [20], and Axiomatic Design Theory [21]. Traditionally, the formulation of functional requirements is mainly dictated by expert designers based on their subjective experiences. Recently, some efforts have been devoted to employ data science and machine learning to support functional formulation, modelling, and analysis.

Functional formulation is the step where a smart product's context-smartness, network-smartness, and service-smartness are instantiated into corresponding functional requirements. As much as possible, the Independence Axiom of the Axiomatic Design Theory should be followed to ensure that the formulated functional requirements can be fulfilled independently [21]. It should be noted that various data operations included in the data-driven design process should be equally mapped to functional requirements of a smart product.

Functional modelling should follow a set of basic principles such as 'complete' (i.e., no necessary functional requirements are excluded), 'minimum' (i.e., no redundant functional requirements are included), and 'independent' (i.e., various functional requirements can be fulfilled separately) [22]. In particular, based on the Functional Basis for engineering design [23], as much as possible, each functional requirement should be formulated as a combination of function entity and flow entity, which models the transformation of material, energy, and signal.

It is also important to clearly differentiate functional requirements from design constraints. The former serves as means to satisfy customer needs, whereas the latter serves as boundary conditions of realizing functional requirements. Design constraints can be classified into input constraint, system constraint, external constraint, and internal constraint [24]. Some customer needs should be translated into design constraints such as cost, size, weight, safety, privacy, etc. Various design constraints play important roles in constructing the practical context of product usage. For example, the design constraints of a robotic vacuum cleaner in a home environment include noise, weight, size, cost, speed, furniture clearance, floor material, floor colour, stair, wall, floor humidity, and so forth [24]. These design constraints can be extracted from online customer reviews through a hybrid approach that involves both machine intelligence and human expertise.

In summary, the input of this step is a set of customer needs extracted from customer voices, whereas the output is a collection of specific functional requirements and design constraints. The control of this step includes multiple design principles of functional modelling and constraint modelling, and the mechanism of this step includes various data operations in tandem with designer knowledge.

Step (3-a): Synthesize a digital structure based on functional requirements and design constraints

In this step, the intangible functional requirements are mapped to tangible design parameters, which are then synthesized towards an integrated design structure. Such a structural synthesis can be conducted in different fashions such as the layer-by-layer zigzagging process prescribed by the Axiomatic Design Theory or the bottom-up integration prescribed by the Analytical Target Cascading [25]. At this stage, the design structure is synthesized entirely within the digital space using various computer-aided design (CAD) tools and computational tools, hence called the 'digital structure'. Since the cost of constructing a digital structure is notably lower than constructing a physical structure, multiple alternative structures should be synthesized, compared, and cross-examined to arrive at the most promising digital structure. In consideration of context-smartness, the digital structure should be situated within different virtual environments that are also constructed through digital means. In particular, various design constraints obtained from Step (2) should be integrated into the digital environment.

Given data obtained from the digital structure situated in the virtual environment, this step involves a variety of data operations such as data mining and data visualization through which the design couplings between functional requirements and design parameters can be made explicit. Design couplings would determine, to a great extent, an artefact's overall complexity [26]. It should be noted that, on top of those design parameters that are intended to fulfil the functional requirements of a smart product, a data-driven smart product should be equipped with additional design parameters (e.g., sensor, actuator, and edge computer) for the corresponding data operations (e.g., data collection, transmission, and storage).

In summary, given the input of functional requirements and design constraints, the former is mapped to a set of design parameters that are synthesized towards an integrated design structure, whereas the latter is used to construct an interpreted environment for accommodating the design structure. Since the structure and environment are both constructed within the digital space, the outputs of this step are a digital structure and a virtual environment.

Step (3-b): Derive expected behaviours from the digital structure

Next, a selection of expected behaviours is derived from the digital structure. As suggested by the name, expected behaviours represent the expectations of a designer with respect to how the structure of a smart product should behave in order to fulfil the functional requirements and comply with the design constraints. A designer's expectations are framed based on his/her unique interpretation of customer voices as well as market competition (i.e., the expected and actual behaviours of competing products). In other words, different designers may have different expectations. The digital structure should be continuously optimized until its derived behaviours meet designer expectations. Besides, certain expected behaviours may also serve as the metrics for concept evaluation, comparison, and selection.

Traditionally, it involves a high degree of designer's subjectivity to derive expected behaviours. Nowadays, various computational, modelling, and simulation tools can be used to derived expected behaviours from a digital structure entirely within the digital space. For the sake of a smart product's context-smartness, it is necessary to simulate the same set of expected behaviours against different contextual conditions (especially concerning design constraints). In the interest of network-smartness, it is necessary to simulate the expected behaviours that are related to product-product connection, interaction, and collaboration. Even the behaviours of certain services can be simulated through digital means. For example, virtual reality can be used to train human workers in terms of safety and maintenance.

Furthermore, target customers can be engaged to interact with the digital structure, being situated in the virtual environment, through digital means (e.g., gamification, simulation, virtual reality, etc.). In this way, designers can gain reasonable expectations and in-depth understandings of not only product behaviours but also customer behaviours (as represented by product-customer interaction). Data on customer behaviours can be compared against the corresponding customer needs obtained in Step (1) for inconsistency, inadequacy, and contradiction.

Step (4-a): Construct a physical structure based on the digital structure

Next, the digital structure derived in Step (3-a) is transformed into a physical structure. Constructing such a physical structure is naturally bounded by additional constraints from the downstream with respect to, for instance, manufacturing capacity and supply chain. As much as possible, the physical structure should be constructed as a vibrant reflection of the digital structure in the physical world. The physical structure of a smart product should be equipped with necessary components (e.g., sensor and edge processor) for the corresponding data operations (e.g., data collection, transmission, storage, and computation). In practice, a number of rapid prototyping techniques (e.g., 3D printing) can be employed to construct the physical structure at a small scale and in a customized fashion.

Step (4-b): Derive physical behaviours from the physical structure

Similar to Step (3-b), a set of actual behaviours are derived from the physical structure to reveal actual product states in the physical world. This step involves data operations such as data collection, data mining, data storage, and possibly data transmission. Physical behaviours can be derived by means of various measuring and sensing technologies, which fall into the scope of data collection. Besides, it is beneficial to explore and track couplings among physical behaviours, which can be supported by data mining. For a smart product, data on physical behaviours can be stored and computed either locally by means of edge computing or remotely by means of cloud computing. In the latter, it is necessary to enable a smart product to transmit data to and from the cloud. Depending on the context, certain smart products may be equipped with extra capabilities to capture certain physical behaviours of target customers as well as peer products in the physical world. Relevant data can be analysed to make sense customer behaviours and make visible abnormal product-customer interactions.

For the sake of a smart product's context smartness, as much as possible, the physical structure should be extensively tested. The testing can be conducted both at the design stage through design of experiment (DoE) and at the usage stage through the Internet of Things (IoT). A smart product's network-smartness enables it to observe physical behaviours of its peers in the same network. Such an ability makes it possible for a smart product to learn from its peers, with respect to their normal (and abnormal) behaviours as well as best practices, to optimize and even customize its own behaviours. Data mining can be used to investigate the hidden patterns of physical behaviours, and machine learning can be used to predict the evolvement of a physical behaviour against a future event or an unknown context. By enabling multiple products to exchange information within a secure network, the effectiveness of data mining and machine learning can be progressively enhanced. Lastly, data on the actual behaviours of a physical structure can be used to enhance a smart product's service-smartness. Specifically, certain abnormal behaviours can automatically trigger requests for corresponding services. For example, a smart refrigerator's abnormal behaviours concerning energy consumption can and should trigger maintenance, repair, and even recycling services. Through machine learning, such abnormal behaviours can not only be pre-defined by designers but also proactively learned by a smart product without designer supervision.

Step (5-a): Connect digital structure with physical structure

The physical structure developed in Step (4-a) is connected to the digital structure constructed in Step (3-a) in order to exchange data between the physical and digital worlds. The ultimate goal is to achieve a two-way communication, coordination, collaboration, and coevolution between the digital and physical structures. On the one hand, the real-time data collected by the physical structure can be used to optimize the digital structure towards a more realistic, holistic, and reasonable reflection of the physical realities. On the other hand, the simulation data generated by the digital structure can be used to optimize, regulate, and control the physical structure. Such a cyber-physical connectivity lies in the centre of digital twin, which is elaborated in Chap. 8 of this book. This step involves data operations such as data transmission, data storage, data fusion, and data analytics.

Data obtained from the cyber-physical connection can be used to enhance a smart product's network-smartness. A consortium of smart products can be connected not only in the physical world but in the digital world. In some cases, it is even possible to connect a physical product with a digital product/service. The cyber-physical connectivity is also beneficial for a smart product's context-smartness. Certain context (e.g., a dangerous environment with many uncertainties), which is too difficult to (re)create in the physical world, can be conveniently simulated in the digital world. In the ideal scenario, a network of smart products in the physical word, a counterpart network in the digital world, as well as a network of service providers can be merged towards a grand network of networks (i.e., a particular form of system of systems).

Step (5-b): Compare physical behaviour against expected behaviour

The physical behaviours derived from the physical structure are compared against the expected behaviours derived from the digital structure. In other words, the outputs of Step (3-b) and Step (4-b) are cross-examined. By doing so, the purpose is to discover any inconsistency, anomaly, and contradiction that may compromise the fulfilment of functional requirements and the compliance with design constraints. Data mining can be conducted to uncover hidden patterns and diagnose inconsistencies. The comparison result can enable a smart product to continuously adapt its structures and behaviours.

The inconsistency between the expected and physical behaviours may be caused by a variety of reasons. Firstly, since expected behaviours represent the expectation of designers based on their interpretation of customer voices, it is likely that such expectations may be disoriented, if not distorted and biased, in the first place. Secondly, the inconsistency may be caused by the ineffectiveness of relevant digital tools (i.e., CAD, simulation, and modelling) that are used to synthesize the digital structure and derive the expected behaviours. Similarly, the manufacturing systems that are used to produce the physical structure, especially on a massive scale, may introduce inaccuracy, uncertainty, and defect as well. Lastly, since smart products tend to have a high level of customization, which may lead to inconsistency regarding both product behaviour and customer behaviour. In particular, product personalization can be regarded as a particular kind of context-smartness, through which, the product dynamically adapts its behaviours to satisfy uniquely situated customer needs.

Step (6-a): Reformulate functional requirements based on changing customer needs

Different from a non-smart product whose functionalities would remain unchanged throughout its lifecycle, the functionality of a smart product should be continuously reformulated in light of the dynamic changes of customer needs. Some functional requirements may become redundant over time, hence they may be removed in order to preserve design resources. Some functional requirements may continue to be desirable, however their design range and target values may be altered in correspondence to customer needs. Some new functional requirements may be included to fulfil emerging customer needs, for example, new applications (APPs) can be installed to a smartphone in order to fulfil new functional requirements. Compared to the traditional design paradigm that mainly relies on structured survey, interview, and observation to track the changes of customer needs, data-driven design is by nature more efficient, responsive, and cost-effective. Through big data analytics, it is possible to track customer preference, customer journey, and market benchmarking constantly.

A smart product's context-smartness, network-smartness, and service-smartness can facilitate the design decision-making about functional reformulation. Customer needs tend to change in response to the contextual variations of product usage. As such, context-smartness enables designers to contextualize the changes of customer needs to distinguish between a universal trend and a unique outliner. Besides, a product's functionalities and behaviours may be affected by its peer products that co-exist in the same environment (context or network). For example, if two smart appliances in the same smart home are equipped with identical, if not the same, functionalities, it is expectable that customer preference would be affected by such a redundancy. As such, in the interest of sustainability, it would be beneficial to remove certain redundant functions to preserve design resources. A smart product's service-smartness can influence customer satisfaction as well. Due to wear or damage, over time, the system range of a physical structure may gradually deviate from the original design range, when relevant services can be requested to restore the missing functionality.

Step (6-b): Reconfigure digital structure based on changing functional requirements

Throughout a smart product's lifecycle, its digital structure should be iteratively reconfigured. Such a reconfiguration may be triggered by a variety of factors. The changes of functional requirements and design constraints, as elaborated in Step (6-a), will promptly trigger the reconfiguration of the digital structure, which is after all constructed based on the conceptual mappings from functional requirements to design parameters. The reconfiguration can also be triggered by inconsistencies between the expected behaviours and physical behaviours, as revealed in Step (5-b), which is especially true if the inconsistency is caused by the ineffectiveness of digital modelling, simulation, and visualization. Lastly, the contextual information collected by the smart product in the field can be fed back to update the boundary conditions of the digital structure. Such a reconfiguration process can be regarded as an iterative design of the digital structure.

Step (6-c): Reconfigure the physical structure

A key feature of smart product is the ability to adapt its physical structure. It should be noted that the physical structure refers to both hardware and software. Similar to the reconfiguration of the digital structure, the adaptation of the physical structure can be triggered by different conditions. In particular, some smart products should be designed to enable different customers to personalize the physical structure in order to cope with practical contexts. Naturally, the reconfiguration of the digital structure, as explained in Step (6-b), also triggers the reconfiguration of the physical structure. Besides, certain services may cause the physical structure to be reconfigured as well.

Step (7): Document structure data for services

In light of design iterations, certain types of data generated in the above steps should be documented continuously. In practice, two kinds of data can be documented: (1) the process data that replicates different operations in the design process and (2) the product data that indicates a product's digital and physical states. The documentation of design data can also benefit from the recent advances of data science. For example, through cloud computing, different stakeholders involved in a design process can access globally distributed data across institutional and geographic borders. Through blockchain, a decentralized consensus can be established with respect to the ownership of intellectual properties generated in a design process. Through federated learning (i.e., a particular paradigm of distributed machine learning), raw data can be stored and computed locally to protect data privacy. Some design data can be documented and shared with customers to make visible a complex customization process, based on which, to justify the pricing of a customized product. Certain physical behaviours should be documented for the purposes of anomaly detection, pattern recognition, and informed diagnosis through data mining and machine learning. Some service data should be documented to reflect changes in a product's whole lifecycle for the purposes of product sharing, remanufacturing, and recycling.

Figure 3 summarizes the step-by-step process of data-driven smart product design. The specific data operations and enabling technologies involved in each step of the process are elaborated as well.

4   Conclusion

This chapter focuses on the data-driven design of a particular kind of artefact—smart product. In sharp contrast to those non-smart products, modern smart products are characterized by strong context-smartness, network-smartness, and service-smartness. The design of smart product can benefit greatly from data science and machine learning, not only because context, network, and service can all be constructed, represented, and reflected through data, but also because smart product by itself constitutes a dedicated data 'maker' that continuously produces new data throughout its lifecycle. Based on the situated Function-Behavior-Structure (FBS) ontology, a systematic design process is prescribed to design smart products with stronger context-smartness, network-smartness, and service-smartness. A number of relevant data operations (e.g., data collection, transmission, mining, visualization, etc.) are incorporated into the process. The application of data-driven design to smart products can not only inform design decision-making but also enhance the internal intelligence of smart products.

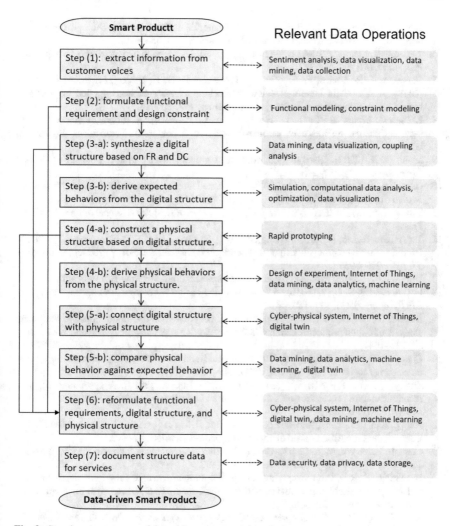

Fig. 3 Step-by-step process of data-driven smart product design

References

1. Porter, M. E., & Heppelmann, J. E. (2014). How smart, connected products are transforming competition. *Harvard Business Review, 92*(11), 64–88.
2. Liu, Y., Zhang, Y., Ren, S., Yang, M., Wang, Y., & Huisingh, D. (2020). How can smart technologies contribute to sustainable product lifecycle management?. *Journal of Cleaner Production, 249*, 119423.
3. Tomiyama, T., Lutters, E., Stark, R., & Abramovici, M. (2019). Development capabilities for smart products. *CIRP Annals, 68*(2), 727–750.
4. Abramovici, M., Göbel, J. C., & Savarino, P. (2017). Reconfiguration of smart products during their use phase based on virtual product twins. *CIRP Annals, 66*(1), 165–168.

5. Liu, A., Teo, I., Chen, D., Lu, S., Wuest, T., Zhang, Z., & Tao, F. (2019). Biologically inspired design of context-aware smart products. *Engineering, 5*(4), 637–645.
6. Tan, K., & Liu, A. (2019). Biologically inspired design of network-aware smart product. *Procedia CIRP, 83*, 767–772.
7. Carvalho, T. P., Soares, F. A., Vita, R., Francisco, R. D. P., Basto, J. P., & Alcalá, S. G. (2019). A systematic literature review of machine learning methods applied to predictive maintenance. *Computers & Industrial Engineering, 137*, 106024.
8. Zheng, P., Wang, Z., Chen, C. H., & Khoo, L. P. (2019). A survey of smart product-service systems: Key aspects, challenges and future perspectives. *Advanced engineering informatics, 42*, 100973.
9. Mont, O. K. (2002). Clarifying the concept of product–service system. *Journal of Cleaner Production, 10*(3), 237–245.
10. Gero, J. S., & Kannengiesser, U. (2004). The situated function–behaviour–structure framework. *Design Studies, 25*(4), 373–391.
11. Tao, F., Liu, A., Hu, T., & Nee, A. Y. C. (2020). *Digital twin driven smart design.* Academic Press.
12. Tao, F., Sui, F., Liu, A., Qi, Q., Zhang, M., Song, B., Guo, Z., Lu, S. C. Y., & Nee, A. Y. (2019). Digital twin-driven product design framework. *International Journal of Production Research, 57*(12), 3935–3953
13. Ulrich, K. T. (2003). *Product design and development.* Tata McGraw-Hill Education.
14. Liu, A., & Lu, S. C. Y. (2016). A crowdsourcing design framework for concept generation. *CIRP Annals, 65*(1), 177–180.
15. Ireland, R., & Liu, A. (2018). Application of data analytics for product design: Sentiment analysis of online product reviews. *CIRP Journal of Manufacturing Science and Technology, 23*, 128–144.
16. Chen, D., Zhang, D., & Liu, A. (2019). Intelligent Kano classification of product features based on customer reviews. *CIRP Annals, 68*(1), 149–152.
17. Zhang, Z., Liu, L., Wei, W., Tao, F., Li, T., & Liu, A. (2017). A systematic function recommendation process for data-driven product and service design. *Journal of Mechanical Design, 139*(11).
18. Xu, Q., Jiao, R. J., Yang, X., Helander, M., Khalid, H. M., & Opperud, A. (2009). An analytical Kano model for customer need analysis. *Design Studies, 30*(1), 87–110.
19. Chan, L. K., & Wu, M. L. (2002). Quality function deployment: A literature review. *European Journal of Operational Research, 143*(3), 463–497.
20. Browning, T. R. (2001). Applying the design structure matrix to system decomposition and integration problems: A review and new directions. *IEEE Transactions on Engineering management, 48*(3), 292–306.
21. Suh, N. P. (2001). *Axiomatic design: Advances and applications.* USA: Oxford University Press.
22. Liu, A., & Lu, S. (2020). Functional design framework for innovative design thinking in product development. *CIRP Journal of Manufacturing Science and Technology, 30*, 105–117.
23. Hirtz, J., Stone, R. B., McAdams, D. A., Szykman, S., & Wood, K. L. (2002). A functional basis for engineering design: Reconciling and evolving previous efforts. *Research in Engineering Design, 13*(2), 65–82.
24. Liu, A., Wang, Y., Teo, I., & Lu, S. (2019). Constraint management for concept ideation in conceptual design. *CIRP Journal of Manufacturing Science and Technology, 24*, 35–48.
25. Liu, A., & Lu, S. C. Y. (2014). Alternation of analysis and synthesis for concept generation. *CIRP Annals, 63*(1), 177–180.

26. Lu, S. C. Y., & Suh, N. P. (2009). Complexity in design of technical systems. *CIRP annals,* *58*(1), 157–160.

Chapter 7
Data-Driven Smart Product Service System

Abstract Due to the rapid development of cyber-physical systems and smart products, a new paradigm of smart product-service systems (SPSS) has emerged. The emergence of SPSS has transformed the way manufacturers deliver value to their product users, as Internet-based services enable products to be constantly updated during their use phase. Furthermore, enhancing product functions during the use stage to better meet users' needs offers revolutionary potential to boost user satisfaction. The successful implementation of SPSS lies in manufacturers' capabilities to anticipate and fulfil users' real-time and contextual requirements by analysing heterogeneous forms of user data. This chapter proposes a framework to illustrate the development process and implementation of SPSS. An illustrative example of a robot vacuum cleaner (RVC) product-service design is used to assess the proposed framework's utility.

Keywords Product-service systems · Smart products · Smart design

1 Introduction

Over the past two decades, changes in the global market have forced manufacturers to shift their business strategies from product-oriented to a service-oriented approach, known as a product-service system (PSS) [1]. As a result, companies now offer various intangible services and physical products to meet diversified customer needs [2]. More recently, the increasing popularity of smart products (SP) has revolutionized the PSS into a smart product-service system (SPSS). SPSS uses information and communication technology (ICT) of SP to connect, collect, and process product information, which is then used to upgrade product services remotely throughout the product lifecycle [3].

SPSS' successful implementation relies on manufacturers' ability to anticipate and fulfil users' needs in real time. Traditional methods to elicit requirements, such as surveys, interviews, and focus groups, are relatively time-consuming and labour-intensive. Due to the slow implementation of these methods, they lack the efficiency needed for real-time SPSS design. In this big data era, the collection, processing, and analysis of large data quantities, known as big data analytics, have fast become a

© The Author(s), under exclusive license to Springer Nature Switzerland AG 2022
A. Liu et al., *Data-Driven Engineering Design*,
https://doi.org/10.1007/978-3-030-88181-8_7

vital technology in SPSS design. Information and communication technology (ICT) can process and transmit large amounts of customer data in real time. All those technologies provide opportunities for analysing users' real-time needs. However, given the unstructured and constantly changing nature of big data, extracting simple and useful information for designers poses a significant challenge.

Against the background, this chapter proposes a data-driven SPSS framework to guide the product-service bundle design process for SP. The framework contains four modules: context detection module, recommendation module, adaptation module, and evaluation module. The rest of this chapter is organized as follows: Sect. 2 discusses the related theories of SPSS. Next, Sect. 3 discusses the data sources for SPSS design. Following this, Sect. 4 outlines the enabling technologies used in the proposed framework, including big data analytics, recommendation systems, and autonomous systems. Then, Sect. 5 discussed the overall framework of the data-driven SPSS framework. Finally, in Sect. 6, an illustrative example of a robot vacuum cleaner (RVC) is given to validate the feasibility and advantages of the proposed framework. Furthermore, Sect. 7 concludes the main contributions of the research, highlighting its limitations and future work.

2   Related Theories

2.1   Product-Service Systems (PSS)

PSS refers to a specific business strategy that combines products and services as a single solution to customers [4]. Compared to conventional business models that solely deliver tangible products to customers, PSS can fulfil customer needs with various services. In addition, PSS' implementation shifts companies' focus on how products are used rather than simply the product itself. The benefits of PSS manifest through two key differences from standard product design. First, in PSS, product improvements are made by upgrading services in addition to the product itself. Second, PSS facilitates the transmission of customer voices from the use phase back to the design and manufacturing phases, enabling manufacturers to have early insights into emerging user needs. As a result, manufacturers will become more responsive to changing markets and more innovation opportunities.

Generally, there are three types of PSS acknowledged in current research: product-oriented PSS, use-oriented PSS, and result-oriented PSS [5]. In product-oriented PSS, companies sell tangible products and transfer product ownership to customers once they are purchased. Companies also need to offer additional services to guarantee the product's functionality and quality. Product-oriented PSS often indicates the use of conventional service. An example of this includes purchasing an automotive vehicle where once the product is sold, limited reparation and maintenance services are offered. In the use-oriented PSS, manufacturers offer product and service bundles through leasing agreements and do not sell the physical product. Instead, manufacturers maintain ownership of the product and are responsible for guaranteeing its usability. An example of this is engine leasing in the aviation industry, where a turbine manufacturer provides and maintains the product for an agreed-upon fee. Finally, in result-oriented PSS, companies provide the customer with a particular result or outcome instead of specific products or service. In this model, the result is agreed upon without defining the physical products used.

2.2 Smart Products (SP)

In recent years, the rapid development of information communication technologies (ICT) such as microchips, software, and sensors has led to the proliferation of smart products (SP). SPs are products that communicate and interact with surrounding environments and other smart products through Internet-based communication services [6]. As shown in Fig. 1, compared with traditional products, SPs possess several technical features, such as intelligence, context-awareness, and reconfiguration [7].

Intelligence refers to an SP's ability to process information and make decisions without human intervention [8]. It is an essential SP feature as it provides the basis for other technical features. An SP's intelligence is formed through embedded high-performance hardware systems, such as microchips and sensors, and analytic capabilities such as artificial intelligence and big data analytics. These technologies enable SPs to process data, make decisions, and improve themselves. The extent of an SP's intelligence is reflected by three characteristics [7]. The first is being able to interact with users. For example, SPs can inform users with product changes, provide user guidance, and prevent user error. The second is that intelligent control goes beyond conventional feedback control, which enables SPs to cope with large amounts of

Fig. 1 Technical features of SP

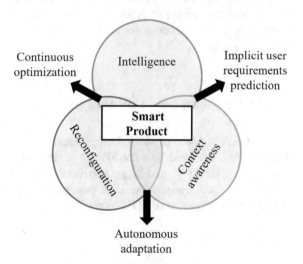

data in order to control more complex activities. The third aspect is autonomy and self-adaptiveness when performing functions, which means SPs can substitute for users to perform functions.

Context refers to situations that affect user preferences and decision criteria. Context-awareness involves an SPs' ability to perceive, interpret, learn, and integrate contextual information to guide decision-making, extend functions, adjust behaviours, and adapt structures [7]. Various equipped sensors enable an SP to collect and analyse data in real time. This information allows designers to understand user activities, surrounding environments, and product performance, drawing links between contextual influences and user satisfaction. Therefore, the context-awareness feature of SP stimulates opportunities for product and service innovations.

Reconfiguration refers to the modification of existing product modules to meet new requirements [6]. SPs typically contain many software modules which are updated remotely. This provides products with the adaptability to change their functions during the use phase of product lifecycle. Therefore, SP offers new potential for performing PSS through reconfiguring software modules and Internet-based services. SP software reconfiguration is categorized into general technical improvements and individual runtime upgrades. General technological improvements refer to the product upgrade for the entire product classes; a typical example is software system updates for smartphones. Individual runtime upgrades provide additional functionalities or Internet-based services, such as parking assisting functions for particular smart vehicles.

SP's unique technical features possess the revolutionary capability to perceive implicit customer needs and deliver breakthrough innovations in customer service [9, 14]. Incorporating SP into PSS, spawning a new paradigm of PSS in this big data era, which is called the smart product-service system (SPSS). SPSS refers to a PSS that integrates SP and Internet-based services to satisfy the requirements of individual users [3]. Furthermore, as SPs depend on Internet-based services, they have access to various modular services that users can select depending on their individual needs. For example, smartphone users can download cloud storage softwares to expand storage capacity.

3   Data Sources for SPSS Design

Data is the foundation of SPSS. Using data from heterogeneous sources, the SPSS extracts useful information for SPs to predict users' implicit needs. Generally, data sources for SPSS design are classified into user-generated content, product-sensed data, and Internet information data. Each of these data sources is defined and explained in the context of SPSS below.

3.1 User-Generated Content (UGC)

User-generated content (UGC) refers to online content posted by web users on social media, e-commerce websites, and discussion forums. There are several notable reasons for the importance of UGC for customer need elicitation. Firstly, UGC is generated by proactive customers themselves and does not require the designer-led traditional approach of pre-defined questionnaires. The advantage of this customer-generated approach is that insights are far less influenced by or confined to designers' hypotheses. They contain unbiased customer voices and produce more accurate insight into what customer needs are most important. Secondly, UGC includes valuable information on many customers from various demographics without the costs traditional surveys would incur. UGC consists of thousands of online reviews which are constantly updated and freely available. Rapid developments in artificial intelligence (AI) technology mean that the large, unstructured, and constantly evolving UGC data can be processed and analysed to produce valuable insights. For example, natural language processing (NLP) software can help extract product features and quantify customers' sentiment. Moreover, computer vision technologies can help designers recognize the composition of the product surrounding environment. As a result, UGC is a promising data source for context modelling and customer need elicitation.

3.2 Product-Sensed Data

Product-sensed data refers to data collected from products' embedded sensor modules. Today, SPs are often equipped with multiple sensors, allowing them to detect and respond to environment changes. Table 1 lists some examples of sensors commonly used by SPs.

Sensor data is used to detect the surrounding environment containing valuable information on the usage contexts of products. By understanding more about how customers use products, designers can develop functions that boost customer satisfaction. For example, smartphones automatically adjust their screen brightness by detecting an environment's light level. Additionally, user interaction data generated from how users interact with a product's interface also provide important insight into customer behaviour. A typical example of this is a smartphone monitoring the frequency and duration users spend on different software applications. Finally, understanding which apps and even specific features within them pique customers' interest provides invaluable insight into their wants and needs.

Table 1 The list of commercial sensors used by SP

Sensors	Functions
Accelerometer	Detect acceleration, vibration, movement
Gyroscope	Exact orientation along three axes
Magnetometer	Detect magnetic field
GPS unit	Determine the precise location
Proximity sensor	Detect how close an external object to the product
Ambient light sensor	Detect the lighting levels
Sound sensor	Measure the loudness of sound
Barometer	Detect weather change and altitude
Camera	Detect objects around the product
Radar	Determine the position of objects around the product

3.3   Internet Information Data

Rapid developments in information communication technology (ICT) allow SPs to connect and share data via the Internet. As a result, SP can receive valuable information from the Internet to help them better serve customers. Generally, Internet information data for SPs is categorized as either product-to-product and website-to-product data. Product-to-product data is sourced from other devices such as personal devices or public infrastructure. For example, vehicles that communicate with each other and road devices that monitor traffic congestion. Website-to-product data is derived directly from Internet sources. For example, smartphones can automatically search for surrounding restaurants and recommend the most popular ones to users.

4   Enabling Technologies for SPSS Design

Faced with a large amount of multi-source heterogeneous data, SPSS requires several technologies for processing and extracting useful information from large quantities of heterogeneous data. This section introduces three critical enabling technologies to realize SPSS, including big data analytics, recommender systems, and autonomous systems.

Table 2 Types of big data analytics

Name	Purpose
Descriptive analytics	To explain what happened from data
Diagnostic analytics	To explain why something could happen
Predictive analytics	To suggest what will likely occur in the future
Prescriptive analytics	To recommend users what to do next to obtain a given result

4.1 Big Data Analytics

Big data analytics is a process of extracting valuable information from extensive datasets using a combination of advanced computational technologies and mathematical techniques. Big data analytics makes it possible to identify customer preferences, market trends, and user behaviours from raw data, which designers use to guide design decisions. Big data analytics is classified into descriptive analytics, diagnostics analytics, predictive analytics, and prescriptive analytics. Table 2 summarizes the purpose of each analytics type.

4.2 Recommender Systems

With the explosive growth of information available online, it is becoming increasingly critical to filter information relevant to users. For example, e-commerce websites contain millions of products and sometimes thousands of the same product. Identifying what information is appropriate in a quick and straightforward process is a complex task. A primary tool for solving this issue is a recommender system (RS), an information filtering system that estimates users' preferences for an item that they may never have looked for themselves. A successful application of RS' is their use in a smartphone application for providing personalized services, such as recommendations for restaurants, films, music, or news. Notably, several design researchers have recently proven RS to be an effective technology to recommend new functions for product and service innovations [10, 11, 12]. For SPSS, integrating RS technology to recommend new functions and services offers further potential for delivering more value to users. Generally, RS are classified into four categories: content based, collaborative, demographic, and hybrid systems. Table 3 summarizes those four types of RS, along with their advantages and disadvantages. Designers must consider the data characteristics, budget, and expected accuracy to select the most suitable RS.

Table 3 Four categories of recommender system

Name	Filtering scheme	Advantages	Disadvantages
Content-based filtering system	Make recommendation based on descriptions of items	Does require data of other users, capable of recommending new items to every customer	Human editors have to tag features of items manually
Collaborative filtering system	Make recommendation to users using information about users' relationship to the item	No need for human involvement, and can make recommendation without any domain knowledge	Items cannot be recommended until items are rated or correlated with other items
Demographic filtering system	Use demographic information such as age, gender, location, etc., for identifying user types	Do not require user history; quick, easy, and straightforward	Recommend items to users of same demographic profiles and results in giving recommendations that are too general
Hybrid recommender system	Mixed approaches that aim to keep the advantage of the combination of methods	Keeps the advantages of the combination of methods	A complex system that is costly to develop

4.3 Autonomous Systems

The autonomous system assists SPs to strengthen and update their functionalities through feature adaptation and reconfiguration. Since the autonomous system can enable SP to deliver appropriate functions to customers in time, it will make SPSS adaptable to constantly surpass customer expectations. An autonomous system identifies contextual information, make decisions, and perform actions in response to external stimuli all on its own [7]. It also contains self-learning capabilities for improving product functions. For example, self-driving smart vehicles are considered autonomous systems as they can learn to optimize hazard perception, and road planning capabilities from human driving behaviours.

5   Data-Driven SPSS Framework

The following section outlines a data-driven SPSS framework for generating a product-service solution for user satisfaction optimization. As shown in Fig. 2, the framework consists of five major modules, namely, the data-driver module, context detection module, recommendation module, adaptation module, and evaluation module. As input, heterogeneous data are collected and processed. Based on

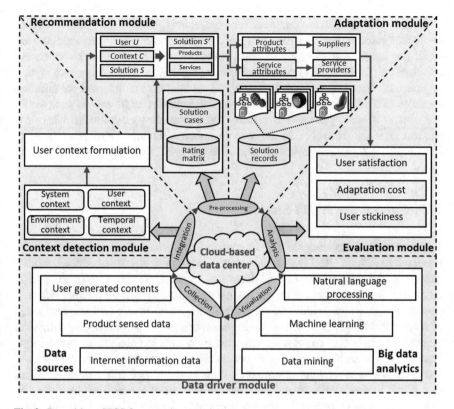

Fig. 2 Data-driven SPSS framework

the information devised from data, the framework can perceive user context, predict user needs, recommend new product-service bundles, adapt product modules, and evaluate the effectiveness of the adaptation.

5.1 Data-Driver Module

The data-driver module consists of data sources, big data analytics, and cloud-based data centre for data processing, analysis, and storage. It is the driving force behind every SPSS module's functionality. Data sources consist of the raw input data from which information is extracted. Examples of data sources for SPSS include UGC, product sensor data, and Internet information data, as outlined above. Big data analytics convert unstructured data into valuable information for decision-making. It involves a five-step process to convert data to information: data collection, data integration, data pre-processing, data analysis, and data visualization [13]. And cloud-based data centre manages data generated from all phases of product lifecycle.

Based on different data modalities, various technologies are used by data analytics to extract information. For example, for text-based data, NLP technologies are used. For image and video-based data, convolutional neural network models are adopted. Furthermore, numerical values can be directly analysed using data mining approaches. Besides, various data analysis techniques are adopted in different modules to extract additional information. For example, UGC data is analysed in the context detection module to identify the surrounding environment where the product operates. However, in the recommendation module, UGC is analysed to recommend product-service solution bundles and predict user satisfaction.

5.2   Context Detection Module

The context detection module derives contextual information. The data sources for this module include UGC (e.g., online customer reviews, discussion forums, professional evaluations), product-sensed data (e.g., location, acceleration, lighting darkness), user interaction data (e.g., use frequency, use duration), and Internet information data (e.g., online communities). By analysing those data, four basic categories of context information can be obtained: system context, user context, environmental context, and temporal context. System context relates to computing system information, such as an IP address, wireless network, and workflow status. User context refers to user demographic information, such as users' age and education. It can also relate to user's tasks (e.g., goals, activities), social connections (e.g., family, friends, colleagues), personal state (e.g., physical information, mental information), and spatial information (e.g., location, movement). Environmental contexts include information about the physical environment where the product operates and is not included by system context and user context; examples include temperature, weather, and surrounding objects. Finally, temporal context encompasses time-related information, such as duration of activity, rhythm, tempo, and stage of life. The frequency at which temporal contexts change varies significantly, with some such as stage of life-changing slowly, while others such as time spent doing an activity change rapidly.

Multiple data analytics techniques are used to extract contextual information from the above data sources. One crucial technique is natural language processing (NLP), which identifies patterns within certain words or groups of words. NLP is often used to identify word types, such as nouns or verbs, and detect patterns between different words. For example, applying entity analysis of natural language processing (NLP) can extract the objects that appear in the surrounding environment where the product operates. Whereas previous research also proves that rule-based NLP can extract user intention of purchasing products [15]. However, although various types of context information can be obtained, building a comprehensive context model is still challenging as most existing data analytics only use a single data source. As a single data source has limited availability, the absence of relevant data will cause the module to fail to detect context. For example, if a context detection module only relies on text-based UGC, the module could not assist users who are not active in generating UGC. Therefore, several data fusion technologies should be adopted to address the data availability issue. Based on previous studies, several data fusion technologies can be applied for context detection, such as Bayes rule-based fusion, fuzzy set theory-based fusion, rough set-based fusion, and so on.

5.3 Recommendation Module

The recommendation module suggests new functions to fulfil the contextual require-ments of target users. In SPSS, various functions are carried by different product-service solution bundles, where users select solution bundles to suit their individual needs. In this circumstance, the recommender system enables SPSS to identify the most feasible solution bundles from historical cases or peer products. All solution history, including any upgrades, changes, and product-service bundle modifications, will be stored. Then, if a similar user context re-occurs in the future, the module predicts what modifications may best suit their needs in different contexts.

For SPSS, a recommender system must possess a context-awareness capability. The context-aware recommendation model is separated into three dimensions: users, contexts, and solutions, with Fig. 3 highlighting their relationship [12]. First, multi-dimensional data (users U, solutions S, context C, customer satisfaction R) are fed into the recommendation model as input data. Second, the recommendation will find the pattern of how U, S, C affects customer satisfaction R. And lastly, given a new input dataset u, s, c, the system will recommend a list of items $\{S_1, S_2, S_3, \ldots, S_n\}$.

In the first step, the user set is denoted by $U = \{u_1, u_2, u_3, \ldots u_{|U|}\}$, the context set is represented by $C = \{c_1, c_2, c_3, \ldots c_{|C|}\}$, and the solution set is denoted by $S = \{s_1, s_2, s_3, \ldots s_{|S|}\}$. The users U can select a solution in set S under different contexts C, and the relationship between U, S, and C can be represented as a three-dimensional space as shown in Fig. 4. In the proposed recommendation model, we assume the relationship between user U and solution S is affected by C, meaning that customer satisfaction about product-service solution bundles varies depending on their context. The three-dimensional space can be decoupled into three two-dimensional matrices: user-context matrix, user-solution matrix, and context-solution matrix. The equation used to represent the three matrices is summarized in Table 4.

After representing the relationship between the user, context, and solution, the next step is to analyse the similarity of any two user-user, context-context, and solution-solution vectors from historical cases with target user, which is formally denoted

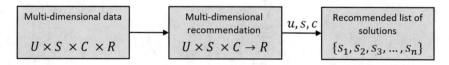

Fig. 3 The procedure of context-aware recommendation

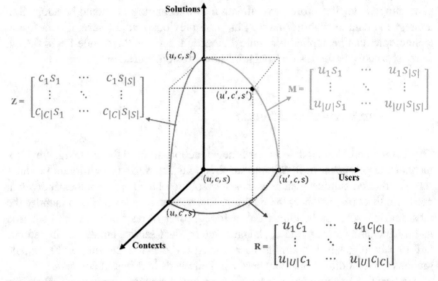

Fig. 4 A representation of the three-dimensional recommendation space on the user, context, and solution

Table 4 Equations of the three matrices

Matrix	Equation	Representation
User-context matrix	$\mathbf{R} = [r_{u,c}]_{|U| \times |C|}$	The number of solutions adopted by the user U_u in context C_c
User-solution matrix	$\mathbf{M} = [m_{u,s}]_{|U| \times |S|}$	The ratings of users U_u to solution S_s
Context-solution matrix	$\mathbf{Z} = [z_{c,s}]_{|z| \times |S|}$	The number of users who adopted the solution S_s in context C_c

as $sim(u_x, u_y)$, $sim(c_x, c_y)$, and $sim(s_x, s_y)$. In our model, the cosine similarity measure (CSM) is used to compute similarity as it has been widely applied in the recommendation model for product and service design. Notably, it has been shown to outperform other measuring methods [10, 11, 12]. Take context-context similarity as an example. The similarity is represented as $\mathbf{A}_{|C| \times |C|} = [a_{c_x,c_y}]$, here the a_{c_x} and a_{c_y} represent the row and column of the two contexts. The $\mathbf{A}$ is calculated using the following equation:

$$a_{c_x,c_y} = sim(c_x, c_y) = \cos(c_x, c_y) = \frac{c_x \cdot c_y}{||c_x|| \cdot ||c_y||}$$

$$= \frac{\sum_{i=1}^{k} a_{c_x,i} \cdot a_{c_y,i}}{\sqrt{\sum_{i=1}^{k} (a_{c_x,i})^2} \cdot \sqrt{\sum_{i=1}^{k} (a_{c_y,i})^2}}$$

Here i represents a particular user and k represents the number of users. The similarity is considered only when the similarity value is among the top k similarities in a given column. The same equation also applies to the solution-solution similarity $\mathbf{B}_{|S| \times |S|}$.

The last step is to build the solution recommendation model for the target user in a specific context. There are two sub-models created, namely, the user-context preference model and the context-solution weight model.

The user-context preference model aims to measure how certain users adopt solutions in contexts that are similar to the target context. This model is denoted as $\mathbf{D}_{|U| \times |C|}$, which can be derived by two matrices $\mathbf{R}$ and $\mathbf{A}$ as follows:

$$\mathbf{D}_{|U| \times |C|} = \overline{\mathbf{R}} \times (\mathbf{A}^k)^T.$$

Here the matrix $\mathbf{A}$ contains $\mathbf{A}^k$ similar contexts, where k is the number of the top solution-solution similarities. $\overline{\mathbf{R}}$ represents the normalized values of $\mathbf{R}$, which is calculated using the following equation:

$$\overline{r}_{u_x, c_y} = \frac{r_{u_x, c_y}}{\sqrt{\sum_{y=1}^{n} \left(r_{u_x, c_y}\right)^2}}$$

The second sub-model, the context-solution weight model $\mathbf{E}_{|C| \times |S|}$, aims to find solutions that have been adopted in a specific context that is similar to a given solution. Similar to the process of developing $\mathbf{D}_{|U| \times |C|}$, the model can be derived from $\mathbf{Z}$ and $\mathbf{B}$, which is represented as follows:

$$\mathbf{E}_{|C| \times |S|} = \overline{\mathbf{Z}} \times (\mathbf{B}^k)^T.$$

Finally, the solution recommendation model is built to do the final estimation of the user-solution matrix which is computed by the product of the two models $\mathbf{D}$ and $\mathbf{E}$. For a given set of contexts in a query q, $q = \{c_1, c_2, \ldots, c_n\}, n \leq |C|$, the relevance score $R(u, c, s)$ of solution S_y for user U_x can be computed as follows:

$$R(u, c, s) = \sum_{c \in q} D_{u,c} \times E_{c,i.}$$

Through this process, the solution recommendation model can predict the value r that a user u possesses with an unknown product-service solution bundle s. Then, based on the ranking list from the highest to the lowest r value, the recommender system will suggest product-service solution bundles to users.

5.4 Adaptation Module

The service adaptation module aims to verify product identity and guide remote product-service upgrades. Adopting an autonomous system can make SPSS self-adaptable to meet the requirement of the individual user in real time. Once the user agrees to upgrade the recommended product-service solution bundle, the system will automatically decompose the solution into various product attribute-related and service attribute-related tasks and assign those tasks to the corresponding manufacturers and service providers. Manufacturers are responsible for designing, producing, and delivering new physical modules to enhance product functionalities and abilities to carry more service modules. And service providers are responsible for meeting the user requirements by authorizing trustworthy Internet-based services or software upgrades.

Whenever a new product-solution bundle is adopted, the detailed product features, attributes, structures, and service components can vary significantly. The adaptation module also manages changes to the SPSS models. After every adaptation, a new solution record is generated to faithfully store the detailed information about product and service changes in the SPSS. This approach has two main advantages. First, the solution record can realize the SPSS management across the product lifecycle as each change is recorded. Second, the solution records provide a data source for the SPSS to further analyse the user's motivations for changing the product-service solution bundle.

5.5 Evaluation Module

The evaluation module's function is to assess whether the solution produced by the recommendation module has effectively fulfilled user requirements. In addition, it helps SPSS verify the solution's cost-efficiency by comparing cost with user satisfaction. Based on the previous study [16], the proposed solutions' effectiveness is measured using the three criteria, namely, (1) maximize the user satisfaction (US); (2) maximize the user stickiness on the SPSS (ST); and (3) minimize the total cost for adoption (TC). Those three criteria are calculated using the following equations:

$$US = \frac{\alpha_1}{|PSB|} \sum_{PSB} (\overline{rate} - \overline{rate_0})$$

$$ST = \alpha_2 \frac{\overline{usageefficiency} - \overline{usageefficiency_0}}{\overline{usageefficiency_0}}$$

$$TC = \alpha_3 \sum_{PSB} (C_P + C_S + C_H + C_I)$$

User satisfaction indicates the average improvement of user satisfaction for each product-service bundle (PSB) in the recommended solution. The user satisfaction rate in the equation is quantified using sentiment analysis of user-generated contents (UGC). The solution is considered effective only if US after improvement is greater before improvement. The user stickiness (ST) measures the improvement of PSB usage efficiency, such as product use frequency and duration, when the user interacts with the PSB. ST supposes that if the user spends more time on the SPSS after adaptation, the user stickiness increases. The TC is the cost summation of physical resources C_P, service-related activities C_S, required human resources C_H, and intellectual resources C_I. The $\alpha_1 \sim \alpha_3$ in the equations are the normalization coefficients that align the magnitude of US, ST, and TC.

After the evaluation of each criterion, the overall cost-efficiency (CE) of the proposed solution is calculated using the following equation:

$$CE = \frac{w_1 \times US + w_2 \times US}{TC}$$

In the equation, w_1 and w_2 are the weighting factors in determining the importance of each criterion. The weighting factors are adjustable based on user preferences.

6 Case Study: Data-Driven SPSS for Robot Vacuum Cleaners (RVC)

This section presents a case study on robot vacuum cleaners (RVC) to illustrate the proposed framework's implementation. The case study analyses common issues with RVC designs and offers potential solutions based on the framework. An RVC is an autonomous vacuum cleaner that cleans dirt and dust from pre-defined floor surface areas. It uses embedded sensors to detect obstacles and hazard within its surrounding environment to measure distances and determine what areas have or have not been cleaned. RVCs can navigate using a floorplan map, where users specify virtual walls, cleaning areas, and no-go zones. Applying data-driven SPSS to RVC offers many advantages to designers and users. For designers, increased data availability can assist in creating more value-added, individualized products and services for future product developments. For users, it can provide real-time reconfiguration of functions and recommendations for how the product will better meet their contextual needs.

The case study detailed below uses 1000 online customer reviews (CR) from five different RVC products. These reviews were downloaded from *Amazon.com* under the 'best sellers in tools & home improvement' category. The most significant context features, product features, service components, and user satisfaction ratings were obtained through several NLP and machine learning techniques. Table 5 shows an example of the analysis result, given a CR content. Here the context extracted is 'large house', the product feature obtained is 'battery', and the user satisfaction rating is '−0.1'.

Table 5 Example of CR analysis

Content	Context	Feature	Satisfaction
I have a larger house, so the battery gets low before it finishes the whole house	Large house	Battery	−0.1

The case study selected the top 6 frequently mentioned context features from CR for context encoding summarized in Table 6. The context of each CR is identified under an encoded scenario $[c_1, c_2, c_3, c_4, c_5, c_6]$ according to the contextual features listed in Table 6. For example, if a user mentions they live in a large house with two inside dogs, two children, and all hardwood floors, their context is encoded as [0, 1, 1, 0, 2, 2]. As those context features involve both qualitative and quantitative data, the encoded context should be normalized. Meanwhile, a total of seven frequently discussed product features and five service components were identified through entity analysis, which is summarized in Table 7.

Aiming to recommend solutions between users, all CRs are encoded for similarity measure. Table 8 shows a small portion of the encoded CR with the context features in Table 6, and the solution lists the current product features and services under the context. The recommendation model is used to compute the context-context similarity and solution-solution similarity among various users.

Table 9 shows an example of the top 4 recommended solutions for a specific user. The user operates the product in a large house with furniture, differing floor surfaces and two pets. The recommendation model suggests the RVC must have 3000 Pa of suction power, 70 mm of height, and 150 min of battery longevity. Also, the component configuring service, such as replacing consumable components such as HUPA filters and brushes, should be provided.

Table 6 Frequently mentioned context feature

C_1: Carpet	C_4: Furniture
C_2: Hard floor	C_5: Pets
C_3: Large house	C_6: Kids

Table 7 Product features and service modules in RVCs

Product features		
P_1: Battery	P_4: HEPA filtration	P_7: Carpet detection
P_2: Suction	P_5: Height	
P_3: Dustbin	P_6: Dirt disposal	
Service modules		
S_1: Parameter configuring	S_3: Voice control	S_5: Software adaptation
S_2: Component configuring	S_4: Maintenance	

Table 8 Examples of encoded scenarios with user satisfaction

CR No.	Encoded context	Encoded product/service solution		Satisfaction in P_2
		Product features	Services	
1	[0, 1, 1, 0, 0.67, 1]	[0.8, 0.4, 1, 0.5, 0.78, 0, 0]	[0, 0, 0, 0, 1]	0.8
2	[1, 1, 1, 1, 0, 0]	[0.8, 0.53, 0.83, 0.5, 0.76, 0, 0]	[0, 1, 0, 0, 1]	0.35
3	[1, 1, 0, 0, 1, 1]	[0.6, 0.45, 0.83, 1, 1, 1, 0]	[1, 1, 1, 1, 1]	−0.13
4	[1, 0, 0, 1, 0.67, 0.5]	[1, 1, 1, 0.5, 0.88, 0, 0]	[1,1, 0, 0, 1]	0.6
5	[1, 1, 1, 1, 0.33, 0]	[0.6, 0.75, 0.83, 1, 0.79, 0, 1]	[1, 1, 1, 1, 1]	0.1

Table 9 Summary of the top 4 recommended functions for a specific context

Context	Recommended solution	Rating
[1, 1, 1, 1, 2, 0]	P_2: 3000 Pa	3.200
	P_5: 70 mm	3.036
	S_2: Component configuring	2.651
	P_1: 150 min	2.032

Once the user accepts the recommended solution, the SPSS decomposes the solution into several sub-tasks, including physical module upgrades and service module upgrades. Then, these sub-tasks are sent to the corresponding manufacturers and service providers. The manufacturer will produce related physical modules to extend the functionalities of RVC. Moreover, the service provider will authorize RVC accesses to the related Internet-based services. Once the adaptation process is complete, the case can be evolved into knowledge with the supervision of domain experts. For example, carpets and pets require RVC to have the higher suction power to clean pet hairs on the carpet. Therefore, once a similar context is identified in the future, the SPSS will automatically recommend the historical solution to new users.

After adaptation, the SPSS should evaluate the cost-efficiency of the recommended solution by comparing the cost of adaptation with improved user satisfaction. In this case study, the cost of adaptation takes the average price of products that possess related features and services in the e-commerce platform. Furthermore, improvements in user satisfaction can be predicted using the deep learning method.

7 Discussion and Conclusion

In conclusion, the increasing popularity of SPs provides manufacturers with a new way of offering product-service solution bundles and deliver value-in-use to customers, giving rise to a new SPSS paradigm. This chapter proposed a data-driven SPSS framework for detecting user context, recommending new solutions, adapting product and service modules, and evaluating the cost-efficiency of adaptation. The proposed framework makes SPSS more intelligent, predictive, and autonomous. The proposed frameworks' main contribution to SPSS includes: (1) the SP is empowered to dynamically fulfil the contextual user requirement; (2) the development process of SPSS is driven by customer data instead of the designer's experience; and (3) the SPSS evaluates the effectiveness and efficiency of feature adaptations, enabling SPs to continually self-improve.

Despite the efficacy of the proposed framework for SPSS, several limitations require further research. First, the case study only uses a single data source to conduct research. Since CR is open-topic data, the context constructed based on CR could vary from the user's real context. Second, due to the limitation of the data source, the modelling context was not comprehensive enough. Finally, analysing CR can help designers obtain the surrounding environment where the product operates; it cannot help designers get user motivation and intention to use the product.

Based on these limitations, there are three key areas for future research to improve this framework: (1) There will be more investigations on how to analyse heterogeneous data to reflect product context, predict user satisfaction, and evaluate adaptation efficiency. (2) A more comprehensive context modelling method will be developed to improve the accuracy of recommendation.

References

1. Qu, M., Yu, S., Chen, D., Chu, J., & Tian, B. (2016). State-of-the-art of design, evaluation, and operation methodologies in product service systems. *Computers in Industry, 77*, 1–14.
2. Wang, Z., Chen, C. H., Zheng, P., Li, X., & Khoo, L. P. (2019). A novel data-driven graph-based requirement elicitation framework in the smart product-service system context. *Advanced engineering informatics, 42*, 100983.
3. Valencia, A., Mugge, R., Schoormans, J., & Schifferstein, H. (2015). The design of smart product-service systems (PSSs): An exploration of design characteristics. *International Journal of Design, 9*(1).
4. Mont, O. K. (2002). Clarifying the concept of product–service system. *Journal of Cleaner Production, 10*(3), 237–245.
5. Reim, W., Parida, V., & Örtqvist, D. (2015). Product-Service Systems (PSS) business models and tactics–a systematic literature review. *Journal of Cleaner Production, 97*, 61–75.
6. Abramovici, M., Göbel, J. C., & Savarino, P. (2017). Reconfiguration of smart products during their use phase based on virtual product twins. *CIRP Annals, 66*(1), 165–168.
7. Tomiyama, T., Lutters, E., Stark, R., & Abramovici, M. (2019). Development capabilities for smart products. *CIRP Annals, 68*(2), 727–750.
8. Liu, A., Teo, I., Chen, D., Lu, S., Wuest, T., Zhang, Z., & Tao, F. (2019). Biologically inspired design of context-aware smart products. *Engineering, 5*(4), 637–645.
9. Rijsdijk, S. A., & Hultink, E. J. (2009). How today's consumers perceive tomorrow's smart products. *Journal of Product Innovation Management, 26*(1), 24–42.
10. Liu, A., Lu, S., Zhang, Z., Li, T., & Xie, Y. (2017). Function recommender system for product planning and design. *CIRP Annals, 66*(1), 181–184.
11. Zhang, Z., Liu, L., Wei, W., Tao, F., Li, T., & Liu, A. (2017). A systematic function recommendation process for data-driven product and service design. *Journal of Mechanical Design, 139*(11).

12. Alhamid, M. F., Rawashdeh, M., Hossain, M. A., Alelaiwi, A., & El Saddik, A. (2016). Towards context-aware media recommendation based on social tagging. *Journal of Intelligent Information Systems, 46*(3), 499–516.
13. Tao, F., Qi, Q., Liu, A., & Kusiak, A. (2018). Data-driven smart manufacturing. *Journal of Manufacturing Systems, 48*, 157–169.
14. Zheng, P., Liu, Y., Tao, F., Wang, Z., & Chen, C. H. (2019). Smart product-service systems solution design via hybrid crowd sensing approach. *IEEE Access, 7*, 128463–128473.
15. Suryadi, D., & Kim, H. M. (2019). A data-driven approach to product usage context identification from online customer reviews. *Journal of Mechanical Design, 141*(12).
16. Li, X., Wang, Z., Chen, C. H., & Zheng, P. (2021). A data-driven reversible framework for achieving Sustainable Smart product-service systems. *Journal of Cleaner Production, 279*, 123618.

Chapter 8
Digital Twin for Data-Driven Engineering Design

Abstract The realization of data-driven engineering design faces three critical challenges, including a limited amount of data, insufficient virtual model quality, and long design lead time. The emerging information technology of digital twin is expected to play an important role in future engineering design. This chapter explores a novel data-driven engineering design process supported by digital twin. Here digital twin provides three features for solving common issues in data-driven engineering design, including the dynamic and comprehensive data collection, the high-quality virtual modelling, and the synchronization between physical and virtual worlds. Digital twin technology overcomes these challenges by supporting the engineering design process in three unique ways: (1) real-time monitoring and data collection in the physical world; (2) identification, diagnosis and prediction of a product's status; and (3) enhancement of human-machine interactions via the virtual entity. A case study of robot vacuum cleaner is used to illustrate the benefits of digital twin for data-driven engineering design processes.

Keywords Engineering design · Digital twin · Smart manufacturing

1 Introduction

As introduced in previous chapters, data tend to play more important roles in engineering design. Conventional engineering design relied heavily upon designers' experience, cognitions, and knowledge [1]. New technological and mathematical advancements have enabled designers to extract invaluable design data from products and customers. This shift transforms engineering design into a far more analytical and predictable process. Data provides numerical evidence to support design decisions and negate the issues caused by designers' subjectivities and inexperience [2]. A data-driven approach involves the systematic analysis of engineering design that highlights design faults, strengths, and potentials in precise, quantitative terms. It incorporates design factors into mathematical models, which are used to forecast a product's performance and customer feedback. A data-driven approach significantly enhances the decision-making capability of designers.

Many technological advancements gave rise to more digital and adaptable data-driven engineering design processes. In the past three decades, the rapid development of computing technologies enabled and promoted computer-aided design (CAD). CAD provides virtual simulations, virtual modelling, and automatic numerical analysis to improve the efficiency of engineering design. Designers often use CAD as a supplement for physical testing, prototyping and empirical research, as it transfers design factors into digital forms [3]. Developments in artificial intelligence (AI) furthered the use of CAD for predicting design performance. Such advances enabled AI to synthesize engineering design data and predict design feature performance. As AI learns and improves, it gradually undertakes primary design objectives.

Meanwhile, market demands are rapidly changing, with customer preferences adapting to more specific needs. New design strategies aim to meet customers' demands; however, the level of personalization and customization expected exceeds the capabilities of most designers [4]. Therefore, designers are increasingly relying on insights derived from customer and product data to adapt to complex and evolving market demands.

The emerging information technology, digital twin (DT), will play an essential role in data-driven engineering design. Glaessgen and Stargel defined DT as an integrated multi-scale, multi-physics, probabilistic simulation to mirror the lifecycle of its physical twin [5]. Here DT was proposed as a three-dimension model, including (1) the physical product and its physical working environments; (2) the virtual world that mirrors the physical world in the digital form; and (3) the data exchange between the physical world and the virtual world. In recent years, Tao's research team enriched DT into a five-dimension model with the additional 'services' and 'connections' [6]. To date, DT has been being enriched with more methods, concepts, and enabling technologies. Compared to other information technologies, DT has three important features: (1) DT achieves the synchronization between a physical entity (PE) and a virtual entity (VE); (2) DT comprehensively collects data in real time; and (3) DT significantly improves the quality of virtual modelling [7]. As the concept of DT is realized, it will become a core tenet of data-driven engineering design.

The following chapter introduces DT and its potential for data-driven engineering design. Firstly, Sect. 2 specifies the enabling technologies of DT and how it works. Secondly, Sect. 3 discusses the significance of DT to data-driven engineering design. Thirdly, Sect. 4 proposes how DT is implemented into the design process. Fourthly, Sect. 5 presents a case study of robot vacuum cleaner to illustrate how DT is incorporated into engineering design. Finally, Sect. 6 concludes this chapter.

2   Introduction to Digital Twin

2.1   Compositions of Digital Twin

The discussion on DT in this chapter is based on Tao's five-dimension model, which consists of a physical entity (PE), a virtual entity (VE), a twin data centre (TDC), services, and connections [6].

PE represents all embodiments, rules, and kinematic mechanisms in the physical world (mainly about the physical product and its working environments). PE is the terminal that executes orders (from designers, peer products, TDC, etc.) and interacts with the physical world. Importantly, PE is the foundation of DT as all other compositions are dependent on it. PE has three main objectives. Firstly, PE undertakes all data collections in the physical world. Secondly, PE regulates physical product to adjust its working status and behaviours based on received commands. Finally, PE forwards data to VE and guides virtual modellings.

VE represents the digital simulation of the product and its working environment and consists of two bidirectional interactions with PE. The first includes the digital embodiments, rules, mechanisms, and behaviours of PE. The second involves the virtual simulations that guide the actions of PE [8]. The VE should represent a completely accurate digital copy of PE in theory. However, in practice, multiple constraints such as computational processing power, data collection quality, and budgets force designers to compensate fidelity for feasibility. Virtual modelling in PE is classified into geometrical, physical, behavioural, and rule categories [6].

TDC represents the platform that converges and processes all data in DT. Internally, TDC contains data from PE, VE, and services. Externally, TDC contains data from designers, customers, working environments, peer products, and online datasets. The large quantity of data processed by TDC supports engineering design procedures, such as concept generation, customization, and service design. In addition to using DT data for immediate design and monitoring of product function, it also provides invaluable information for future DT development. For example, historical product data enables the VE to build and predict the behaviour of future models. This is achieved through advanced data analysis methods, which allow trend formulation, risk management, and working evaluations to occur. TDCs also interact with each other, exchanging data to enhance a product's performance, and enhance the DT's self-learning capability [6].

Services of DT represent functions that satisfy internal and external demands. Functional services (FS) support represent those support work of DT, such as 'fuse data collected from peer products' and 'assemble virtual models'. Business services (BS) are those which satisfy external demands from users or customers, such as 'risk management for production line' and 'diagnose fault of manufacturing failures'. FS and BS are determinative of each other. FS is required to achieve BS, while BS guide FS development. Management, optimizations, and usages of services rely on a software platform to improve quality [6].

Connections of DT are defined as interactions among each composition (PE, VE, TDC, and services platform). A connection system's main objective is ensuring accurate data transmission among compositions. They also help convert data forms, enhance data security, and data retrieval speeds. When building connections in DT, designers must confront two challenges. Firstly, data receivers must retrieve data from multiple sources concurrently. However, many issues arise when the data from these sources exist in different forms, requiring unique transfer protocols and transmission methods. Processing this variation of data forms risks distorting data and creating errors [9]. Secondly, the physical distance between VEs and PEs makes synchronization issues, as transmitting and receiving large data quantities over long distances is limited by existing communication technologies.

2.2   Enabling Technologies of Digital Twin

The existence of DT depends on a variety of enabling technologies, with Fig. 1 providing some examples. The enabling technologies for PE mainly involve data identification and measurement functions in the physical world. In this PE domain, the sensor network provides the foundation for all data collection. Generally, sensors are designed to measure individual data categories, though sometimes multiples sensors each measure a single category to increase accuracy. The network of sensors must perform measurements and wirelessly transmit information to the TDC. For example, sensors may measure physical attributes such as volume, weight, and Young's modulus [10]. Even physical phenomenon such as cracks, fractures, and crystal status are measurable by laser and image recognition technologies [11].

Enabling technologies of VE achieve virtual modellings. As aforementioned, virtual modellings of VE are classified into geometry modellings, physical modellings, behaviour modellings, and rule modellings. In terms of geometry modellings, the wireframe modelling and solid modelling effectively support the formulation of geometrical data [12]. Physical modellings rely on material attributes that can be input from TDC, peer products, and online datasets. Behaviour modelling can be achieved by finite state machines and activity models. The former indicates the static status while the latter describes the dynamic status. Rule modellings require the rule extraction and rule association. Researchers proposed linear support vector machines and SQL cursor interface for the rule extraction and association, respectively [13, 14].

Enabling technologies of TDC undertake data management and data processing. The data storage and data fusion are the two main objectives of data management [5]. NoSQL is one popular approach of data storage and extraction [15]. In terms of data fusion, it can be achieved by random methods and artificial intelligence. In addition, data fusion is generally classified into raw level, feature level, and decision level. Random methods can be applied in all condition while artificial intelligence cannot

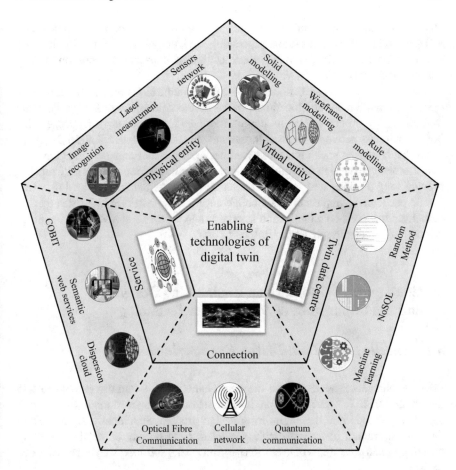

Fig. 1 Examples of digital twin enabling technologies

be applied in the raw-level fusion [16]. In terms of the data processing, interactive spatial data analysis is the most common approach [17]. Besides, machine learning enables abilities of decision-makings of TDC.

Enabling technologies of services cover service generation, service management, and on-demand use of service [18]. Service generation focuses on the perception and encapsulation of service resources. For these purposes, middleware and dispersion cloud can schedule resources to match users' requirements to support resource perceptions, while semantic web services and service-oriented architecture incorporate different components of DT into usable applications to support resource encapsulations [18]. Service management consists of searching, matching, and scheduling of services. Information system platform technologies like ITIL, COBIT, and Six Sigma undertake service deliveries, business alignment, and implementation organization to achieve service management. On-demand use of service represents the management of business and transactions. Emerging transaction technologies such as blockchain, online transaction processing, and interorganizational information system improve localization and personalization of transactions while optimizing business securities [18].

The enabling technologies upon which DT communications depend on a product's operating environment and location. Optical fibre communication is a typical local transmission approach due to its low interference level and quick response time. Various wireless networks are used for remote operation, such as 5G and 4G cellular networks. However, such networks have interference issues and are limited in their data transfer capacity. In the future, quantum communication methods may reliably provide long-distance and secure transmission of large data quantities [18]. This can be particularly important when collaborating manufacturing industries located globally. Moreover, communication standards and protocols are essential to ensure smooth data transmission among different ports.

3   Challenges of Data-Driven Engineering Design and the Importance of Digital Twin Technology

3.1   Challenges of Data-Driven Engineering Design

Previous chapters have explained the many benefits of a data-driven engineering design approach. However, there are several major challenges to incorporating data analytics methods into the engineering design process. Specifically, these include the limited amount of data available, the insufficient fidelity of virtual modelling techniques, and constraints of leads times in processing and analysing large quantities of data.

Despite endless possibilities for collecting new forms of data in the engineering design process, the ability to collect and process this into useful information is limited. There is a high economic cost involved in the technology and human labour required to analyse large quantities of design data. For example, extracting useful insights from online product review data not only requires advanced API and data processing technologies but highly skilled knowledge to manipulate and interpret the data. Consequently, this process is expensive and time-consuming. Conventional design methods of data collection were far simpler yet focused on few areas of product improvement. For example, if a designer wanted to evaluate the power consumption of a refrigerator, they could monitor this single variable over its lifecycle and under different environments. The following analysis of the data was far more straightforward, as it was used solely to investigate a single variable, not discover unknown information across multiple variables. Therefore, designers must compromise between the cost and time involved in analysing large amounts of data via advanced methods and the finite scope of analysing smaller amounts of data with faster, simpler ways.

Although virtual modelling techniques continue to improve, their quality is yet to resemble a physical model. Designers are increasingly reliant on virtual models for design verification and decision-making. Models are beneficial for this process; however, they do not possess the fidelity necessary to predict the exact behaviours of a physical product. Therefore, designers must understand what functional gaps exist between virtual and physical product performances. Conventional virtual models cannot evaluate all design factors. They are effective for design configurations and dynamic behaviours but struggle to replicate complex or unpredictable factors such as the operating environment.

The final key issue with data-driven processes in engineering design is the additional lead times and complexity required for large-scale data analysis. A significant amount of time is needed to collect comprehensive datasets and process these large amounts of data into useful information. However, high levels of market competition require shorter lead times to release new products. Designers must therefore examine the trade-off between enhanced data-driven design and lead times. For example, if a designer wanted to improve the aerodynamic performance of a wing, a range of time-consuming steps are required to produce useful data. These include measuring the wing's dimensions, inputting these dimensions into a virtual model, inputting environmental parameters, running a simulation, and evaluating the results. The cost-benefit of this data-driven approach versus conventional approaches must be considered for various design factors and optimized accordingly.

3.2 The Case for Digital Twin

Compared to other information technologies involved in engineering design, DT has three special features which offer transformational benefits to engineering design. These include: (1) dynamic and comprehensive data collection; (2) enhanced fidelity of virtual models; and (3) enables a synchronization between virtual and physical products. The potential and advantages of DT for data-driven engineering design are explained below.

DT provides an entirely new approach for collecting and analysing large amounts of data from a wide variety of sources. DT simultaneously collects and analyses data from the physical world, the virtual models, and outside sources such as online sales and reviews. Using a wide variety of sensors, data from the physical world is continuously fed into and used to enhance the virtual model.

Furthermore, high-quality modelling generated by DT enables significant improvement in design verification. In theory, the virtual model will eventually mirror a product's performance and working environment. Such high-fidelity modelling is achievable via a constant feedback loop between the virtual and physical realms, with new data constantly enhancing the self-learning capability of designs.

Finally, DT significantly reduces the lead time of data-driven engineering design. In principle, DT autonomously collects and uses machine learning methods to improve constantly. Advanced mathematical methods, computational power, and controlling the scope of analysis enable the virtual model to self-learn and improve. Recording, processing, transmitting, and analysing the data become autonomous, eliminating the long times typically required for such data analysis methods. Using this information, more efficient processes for producing new virtual models are possible, further reducing the lead time required to test new virtual designs.

4   Digital Twin for Data-Driven Engineering Design

The following section discusses how DT services support data-driven engineering design. As shown in Fig. 2, DT provides three main services, including: (1) real-time monitoring and data collection in the physical world; (2) identification, diagnoses, and predictions of product functionality; and (3) enhancement of human-machine interactions via the VE.

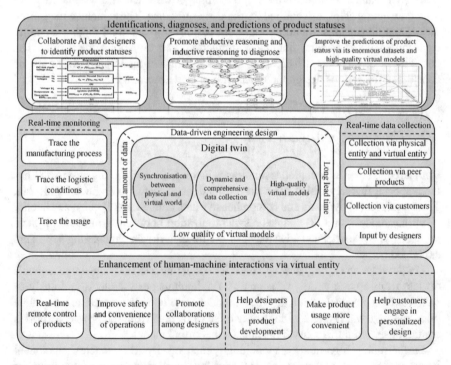

Fig. 2 Services of digital twin for the data-driven engineering design

4.1 Real-Time Monitoring and Data Collection in the Physical World

DT enables the real-time monitoring of physical products via its abundant sensor network and low-latency communication. Conventionally, designers monitor product status by analysing working data or user feedbacks, which generally is only available after the product has been made. With DT, the synchronization between PE and VE allows designers to monitor product performance and environmental conditions via their VEs in real time. DT also enables a product's entire life cycle to be analysed, including processes such as design, manufacture, use cases, maintenance, and disposal via a VE.

DT enables the real-time monitoring of abundant data categories, with examples shown in Table 1. Analysis of product data from customers, design concepts, and operating environments gives designers more precise insights into how their products perform, issues with how they are used, and what the most vital customer needs are. Data are collected via hardware resources, software resources, and network in PE,

Table 1 Examples of data collected by digital twin

Categories of data	Composition of DT	Examples of data
Data of physical products	PE	Stress on the product structure, real-time performances, temperature of the control unit, etc.
Data of working environments	PE	Humidity in the room, air flow speed, signal strength, etc.
Data of virtual models	VE	Dimensions of the frame, coordinate, properties of material, etc.
Data of simulations	VE	Simulated vibration frequencies, acceleration, viscosity, etc.
Data of customers	PE, TDC	Customers' ratings, customers' preferences, customers' demographical data, etc.
Data of concepts	PE, TDC	Popularities of concepts, expected performances of concepts, design tolerances, etc.
Data of peer products	PE, TDC	Resource supply of peer products, dynamic statuses of peer products, design information of peer products, etc.
Data of manufacturing	PE, VE	Material usage, power consumption of manufacturing, lead time of assembly, etc.
Data of maintenance	PE, TDC	Load cycles before fracture, causes of failures, maintenance periods, etc.
Data of disposal	PE, TDC	Recycling rate of components, lifetime of products, failure rate of retired components, etc.

VE, and TDC [18]. In PE, DT collects data of product statuses (idle, energy consumption, power output, fatigue, etc.), product working, environments, and interactive targets. In VE, DT convergences data generated by simulations and virtual modelling. In TDC, DT summarizes data of historical physical tests, virtual simulations, design cases, customer voices, and online resources.

Additional sources of important design data are beyond those collected by the PE, VE, and TDC. One vital source of data is the designer. For example, designers can input data unobtainable via the DT, such as previously unseen or extreme operating conditions. Designers also play an important role in assessing the quality of input data from the PE and VE, such as incorrect readings from the PE sensors or VE simulation errors. Another source of design data is obtained from peer products, such as other smart products used in similar circumstances. The most important design data source is customers, whose insights provide the foundation of design decisions at each stage. For example, during the concept design phase, customer voices guide function and constraint designs. Customer feedback from product usage also provides invaluable insights for future improvements. When designing a DT system, ensuring various product data sources are collected in real time is vital.

4.2 Identifications, Diagnoses, and Predictions of Product Statuses

Foundations of identifications, diagnoses, and predictions are perceptions and descriptions of product statuses, such as working outputs, energy consumptions, and malfunctions (which can either be a reason or conclusion in reasoning). Collected data are too unstructured to reflect product statuses before arranging and packing. Contextual data provides a useful approach to achieve this purpose [19]. Table 2 shows some examples of contextual data. Physical contextual data describe product statuses, social contextual data, user contextual data, and behaviour contextual data. Each contextual data serves as a variable to support different approaches to data analysis. For a typical example, Belief Network (Bayesian Network) is a paradigm of abductive reasoning in computer science to diagnose system diseases. Contributions and conditional dependencies of each variable are computed to analyse probabilities of possible diseases [21]. The amount and categories of contextual data are main constraints that affect identifications, diagnoses, and predictions. Enabled by DT, contextual data can be dynamically collected. Historical contextual data keeps accumulating to expand datasets of DT that can reflect more product statuses. Meanwhile, DT comprehensively collects contextual data from multiple sources that expand categories. Data sources like working environments, peer products, and customer feedback cover a wide range of contextual data supporting identifications, diagnoses, and predictions.

DT unites AI and designers for product status identification. Contextual data collected from PE, VE, and TDC integrate into training datasets that enable the AI

Table 2 Examples of contextual data to describe product statuses

Categories of contextual data	Example
Physical contextual data	Timetable
	Temperature
	Location
	Passenger
	Space layout
Social contextual data	Resource supply/received
	Notification
	Interaction with peer products
	Service output
	Entertainment
User contextual data	Demography
	Mood
	Using habits
	Preferences
	Knowledge of products
Behaviour contextual data	Energy generation
	Command
	Computation
	Movement
	Energy consumption

of TDC to learn features and patterns of product statuses. As the DT collects and accumulates contextual data in real time, AI will keep verifying and reinforcing learning outcomes. With enough iterations DT will be gradually capable of identifying product status autonomously. Designers play a crucial role in helping AI models interpret contextual datasets. For example, designers input the mathematical models used to self-learn and identify patterns. They also evaluate different mathematical models and techniques and perform manual analyses to investigate specific problems. As VEs begin to mirror PEs accurately, designers will observe product statuses remotely.

Through the learning process, AI of DT will gradually be capable of identifying product statuses automatically. In terms of designers, in one aspect, designers can help AI of DT learn contextual datasets. For example, designers can input mathematical models that supervise AI to learn patterns from contextual data. Besides, designers can extract invalid data from datasets that prohibit their interferences to the learning process. In another aspect, designers can directly and intuitively identify products via DT. As DT builds the digital mirror of physical product in VE, designers can observe product statuses from VE remotely in real time. Besides, as PE can be operated with flexibility, designers can control the viewing perspective, zoom, and animation to observe product statuses explicitly.

DT promotes abductive reasoning and inductive reasoning to diagnose. Abductive reasoning is one typical foundation of problem diagnoses. Based on abductive reasoning, designers seek for most likely conclusion (the reason to explain a design problem) from a set of observations. The development of computer science enables AI to automatically diagnose some problems by learning datasets [20]. Meanwhile, inductive reasoning is often applied to diagnoses when there are limited changes in problem situations. Designers and AI make conclusions based on existing cases. For both abductive reasoning and inductive reasoning, the amount and quality of datasets are main constraints that affect diagnoses. Enabled by DT, datasets can be significantly expanded by continuously and comprehensively collecting contextual data, which builds a solid foundation for AI and designers to learn, arrange, derive, and conclude for diagnoses. Besides, DT enables the verification of diagnosis outcomes. Problems and resolution strategies of diagnoses can be simulated in VE via mathematical models or virtual models. Feedbacks of verifications will improve the diagnosis process.

DT improves product status predictions using large datasets and high-quality virtual models. Most information technologies make predictions based on mathematical models generated from collected data. A larger dataset tends to promote quality of predictions. The continuous collection and processing of data enables the dynamic improvement of mathematical models. Eventually, due to the rapid expanding dataset, the quality of virtual model simulations is sufficient for accurate prediction of product physical behaviour. Besides, compared to conventional information technologies, prediction outcomes (physical phenomena, product behaviours, mathematical trend, etc.) of DT can be presented concurrently in numerical forms, in images, and in animations in VE and TDC.

Identifications, diagnoses, and predictions of product statuses support engineering design in multiple ways. For example, identifications of product statuses help designers notice design problems and defects during their real-world use. Diagnosing a product's status assists in identifying design problems at earlier stages of the design process. Finally, predicting a product's status ensures practical evaluation of design decisions and a product's performance and reliability.

4.3   Enhancement of Human-Machine Interactions via the Virtual Entity

Using DT, human-machine interactions in a high-fidelity virtual form offer revolutionary possibilities for engineering design. Conventionally, designers and customers directly interact with physical products. Such human-machine interactions are often constrained by physical environments (high temperature, high voltage, toxic material, etc.), distances, operation complexities, and human disabilities. As DT achieves the synchronization between PE and VE, designers and users can operate physical products via the VE. Engineers develop dedicated application programs to monitor, control, and operate virtual models. When designers and customers control virtual models in application programs, physical products will keep the synchronization with virtual models to be controlled indirectly. Such human-machine interactions via VE have four important advantages: (1) designers and customers can interact with products remotely without the limitations of distance and the physical environment; (2) designers and customers can interact with products during early product development stages; (3) interactions via VE eliminate the limited operational skill and abilities of designers and users when manoeuvring physical products; and (4) the replicability of virtual models enable vast numbers of designers and customers to share one single product.

These advantages of DT have a major impact on how designers develop and use products. For example, real-time remote monitoring of the whole product life cycle significantly enhances the ease with which designers test products at each stage. Modern production processes are highly distributed among separate factories, workshops, laboratories, and offices worldwide. DT eliminates the many geographical barriers which prohibit designers from testing products at different stages. Not only does this provide designers with far greater control over each design stage, but it also allows more designers to participate in the design process. VEs keep virtual records of design changes made by designers at different stages, allowing for far greater collaboration and communication between designers and other stakeholders in the design process. DT also offers significant safety benefits for productions and tests of products. Product development involves a range of potential safety risks, such as harmful particles in workshops, high-temperature testing environments, and hazardous machines and chemicals used in the manufacturing process.

The human-machine interactions achieved through the VE also have notable benefits for customers. When testing products, VEs eliminate many of the issues associated with testing a physical product. For example, the low cost and low risk of harm when using a VE allow customers to engage in more training and test for more complex products. Previously, the cost of training and the risk of damage to physical products impeded thorough testing of some products. The ease by which customers access, test, and provide feedback on designs via the VE exponentially exceeds traditional physical testing methods. Another benefit of the VE is that it allows customers with new opportunities to engage in product development. Often loyal customers are eager to know more about their products and share ideas about improving them. Through the VE, customers can observe and comment on the various stages of product development, offering unique insights for designers while strengthening brand loyalty with customers.

5 Case Study

This section illustrates how DT supports data-driven engineering design using a case study of a robot vacuum cleaner (RVC). The RVC is an ordinary and typical smart home product, which shares many of the technical foundations required for DT. RVCs have sensors, wireless connectivity, and communication to multiple devices and use AI processes to perform their tasks autonomously. The case study below provides a theoretical framework for integrating DT technology into the design process of RVCs.

5.1 The Development of Digital Twin for the Robot Vacuum Cleaner

This section will provide an illustrative example of developing DT for RVC. The development process is separated into PE, VE, TDC, interactions, and connections. As shown in Fig. 3, each part comprises several systems.

For developing the PE, designers must incorporate systems for advanced data collection and transmission. The PE must collect data related to its working environment, such as temperature, humidity and room layout, and the working status of the RVC itself, such as power consumption, suction force, and movement speed. RVC data collection systems require multiple measurement technologies, such as laser measurement, confocal sensing, and hygroscopic measurement.

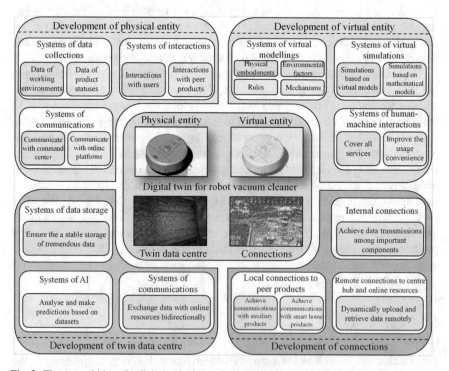

Fig. 3 The general idea of a digital twin for a robot vacuum cleaner

For developing the interaction systems, the DT design must interact with users and peer products. Examples of potential interactions with users include collecting demographic data and discovering links between certain customer behaviours and working situations. Interactions with peer products, such as extracting working statuses or exchange control instructions, also provide designers with valuable insights into how designs meet customer needs. Third, communication systems for DT must handle substantial data transfer quantities. Exchanging data with other smart home products, online customer voices, working situational data from the PE to the VE, and inputs from the VE to the PE requires high bandwidth wireless communication technologies, such as 5G or light wave transmission, or else it risks slow performance or interference.

RVC VEs require advanced virtual modelling, virtual simulation, and human-machine interaction systems. Virtual modelling systems contain comprehensive information about a product's physical properties (dimensional and structural), mechanical properties (motor power, belt-drives, gears, and shafts), electrical properties (electrical power consumption and electronic communications), and environmental factors (airflow, temperature, humidity, light levels). Virtual simulation systems consist of advanced computational technologies and mathematical techniques for predicting the working status of the RVC. Simulations should accurately predict the behaviour of the VE under specific conditions. The human-machine interaction system is the portal for users and designers to control the RVC and its DT. This involves the software interface or smartphone apps users use to program the RVC, and how designers interact with PEs via the VE. When designing a human-machine interaction system, optimizing the balance between technical capability, and ease of use is vital.

When putting together the TDC, designers must incorporate data storage systems, AI technologies, and communication with online platforms (the online cloud hub that store data and collaborate stakeholders). The data storage system must store large amounts of data related to all facets of the DT. It is also here that AI technologies analyse this data to predict working status, environmental influences, and customer demographic behaviour. The TDC both collects data from and uploads to online platforms.

When designing RVC connections, designers must incorporate systems for internal connections within RVC components, between peer products, the TDC, and online platforms. Across all connection systems, fast data transmission speeds, high levels of stability, and rigorous safety requirements are paramount. Internal connections must efficiently transmit information between different components to ensure function, for example, transmitting speed requirements to the motor output and obstacle detection sensors to change the direction of travel. Connections to peer products may include communications between auxiliary products, such as virtual walls, robot mop cleaners and charging hubs, or other smart home products, such as smart home controller surveillance cameras or climate control systems. Connections to the TDC and online platforms require sophisticated telecommunications technology for remote data transfer. In practice, all connection systems are prone to delays and disruptions. Therefore, data transfer and reliability limitations will inevitably obstruct the synchronization between the PE and VE.

5.2   Real-Time Monitoring and Data Collection for the Robot Vacuum Cleaner

DT allows designers to trace RVC manufacturing and usage in real time. Table 3 presents some examples of data collected by the DT of RVC. The emerging smart factory is expected to have three levels of DT: DT of the factory, DT of machines, and DT of products. For example, the PE of an assembly robot can scan, model, and upload data related to a product's assembly. It is also possible for such machines to record other manufacturing data, such as injections, extrusions, castings, and choice of tools. Such data enables designers to identify ways to reduce complexity, material consumption, power consumption, and defect rate.

Moreover, during the usage, the sensors network enables DT to continuously retrieve data of each component to monitor its working status and health conditions. For example, when suctioning debris off the floor, a magnetic sensor records motor RPM, infrared sensors monitor temperatures of moving parts, and a pressure sensor measures air flow speed in the wind tunnel. Monitoring a combination of this data provides a comprehensive overview of an RVC's suction performance.

In addition to the RVC DT, PE data is also obtainable from other peer products with similar working environments to the RVC. For example, air conditioners may record temperature or humidity data, and surveillance cameras may record data of moving objects nearby the RVC or even the speed of the RVC itself. Furthermore, it is also possible for the PE to obtain customer voice-related data. For example, customers may speak directly to the RVC or communicate to the RVC via a smartphone, which provides data relating to customer attitudes or even suggestions for product improvements.

Despite the potential for autonomous data collection and analysis to occur with RVCs, it is still necessary for designers to assess data quality and prediction accuracy. The risk of incorrect data collection always remains due to malfunctioning or damaged sensors, communication interferences or software bugs. Therefore, all DT designs require the designer's knowledge and experience to evaluate data quality and relevance such as identifying data, deleting incorrect data, or troubleshooting to find the source of inaccurate data. Furthermore, there may even be facets of a DT design that are not possible to model accurately. One example of this could be the structural properties of a new composite material, of which no data have been collected as references. It may be impossible to predict its behaviour by analysing datasets, which may require designers to input, modify, and verify its analysis manually.

Table 3 Examples of data collected by the digital twin of robot vacuum cleaner

Categories of data	Composition of DT for collection	Examples of data
Data of the RVC in physical world	PE: The sensors network (magnetic sensor, deflection sensor, temperature sensor, etc.)	Shear stress of the turbine shaft, suction power, temperature of centre processor, etc.
Data of working environments of the RVC	PE: The sensors network	Humidity in the room, illumination in the room, strength of connection to the smartphone, etc.
Data of virtual models of the RVC	VE: The monitoring software of virtual models	Dimensions of the RVC, location of the RVC, properties of RVC materials, etc.
Data of simulations of the RVC	VE: The monitoring software of virtual simulations	Simulated brushing, suction, mopping, etc.
Data of customers of the RVC	PE, TDC: Web crawler, voice recognition, dedicated application program on the smart phone, etc.	Customers' ratings of the RVC, customers' using habits of the RVC, customers' demographical data, etc.
Data of concepts of the RVC design	PE, TDC: datasets of RVCs, peer products, online resources, etc.	Material properties of different RVCs, expected endurance of the battery, the range of suction power, etc.
Data of peer products of the RVC	PE, TDC: online resources of peer products, the sensors network, PE of peer products, etc.	Signal strength of smart home controller, movement speed of robot mopping cleaner, distance between virtual walls, etc.
Data of manufacturing of the RVC	PE, VE: sensors on manufacturing parts, scanners on manufacturing machines, virtual models of production lines, etc.	Material usage, power consumption of manufacturing, lead time of assembly, etc.
Data of maintenance of the RVC	PE, TDC: datasets of parts, sensors on parts, mathematical predictions, etc.	Load cycles of each rotation, frequencies of abnormal vibrations, consumptions of battery endurance, etc.
Data of disposal of the RVC	PE, TDC: datasets of recycling, sensors on parts, sensors on disassembler, etc.	Endurance of rollers, energy consumption of disassembling motors, length of cracks, etc.

5.3 Identification, Diagnosis, and Prediction of a Robot Vacuum Cleaner's Performance

As mentioned above, the identification, diagnosis, and prediction of a RVC's status are based on its contextual data. Contextual data appears in different forms, such as proportion (e.g., 63% of users age between 22 and 35), frequencies (e.g., fully charged battery can support three times quick cleaning), and parameters (e.g., suction power of a RVC is around 22 W). The raw contextual data collected from an RVC's PE, VE, TDC, and online resources are classified and processed to provide useful information for designers.

An RVC's product status is identified by AI verified by designers. Table 4 shows a pair of contextual data which describe an RVC's mid-level cleaning mode and charging mode, respectively. Different RVC statuses show varying contextual data values of the two sets, which can be identified by AI. For example, as shown in Table 4, the RVC mid-level cleaning mode generates 16.63–18.31 W of kinematic power while operating. In comparison, the charging mode generates 3.8–5.4 W electrical power without any kinematic power use. If the DT detects 4–7 W of electrical power being generated and no kinematic power is output, it easily determines that the RVC is in charging mode. Other practical conditions, such as RVC operational speed, require more contextual data forms identifying an RVC's status. Consider an example where the exact RVC status is unknown but where only some operational parameters are being measured. The detection of linear accelerations indicates the RVC is in cleaning mode; however, the kinematic power is unknown. If thermal power is 26.77–29.41 W and suction power 26–31 W, such measurements identify that the RVC is in high-level cleaning mode. Despite the reliability of quantitative, machine learning-driven methods of product status identification, DTs are also important tools for verification by designers. For example, it is possible for a designer to monitor a VE's identifications in various simulations, and evaluate its overall accuracy. If inconsistencies are detected, designers will make corrections, which also serve as feedback to improve analysis accuracy of DT.

An RVC's DT also supports the diagnosis of its status. A typical diagnosis approach uses the Bayesian Network. The diagram in Fig. 4 presents a simple RVC diagnosis example using the Bayesian Network. A user notices abnormal noise of the RVC caused by vibrations of the shield, so the DT of RVC needs to diagnose the cause. It is assumed that the DT has collected contextual data relating to the RVC's vibration frequency, noise level, and airflow speed. The DT first infers the real suction power output (r) and heat level of the shaft on wind turbine (h) to diagnose the problem. Based on contextual data and historical recordings, r and h have two possible conditions, for which probabilities (p) are shown. Then, DT needs to infer debris condensation (c) and wind turbine shaft malfunctions (f) before determining the vibration root cause. Assume that c is inferred based on primary contextual data and r; thus, c have two possible conditions for each r. The p of each c varies. In terms of f, assume that f is affected by r, h, and primary contextual data. In this case, f can have three different conditions with different p values for each combination of r and h. Finally, based on contextual data and historical analysis, potential causes of malfunction (k) have three candidate results for each combination of c and f. After the multiplication and addition of each combination from r and h to f and c, the possibility of each k can be obtained and forwarded to the user (the possibility

Table 4 Examples of contextual data of robot vacuum cleaner

Contextual data	Mid-level cleaning mode	Charging mode
Timetable	55% (the proportion of all collected working cases) 2–5 pm, 45% 8 am – 11 am	60% 9 pm–1 am, 40% 2–7 pm
Temperature	57–65 °C	42–49 °C
Location	27% (the proportion of total working time) in the living room, 18% in the kitchen, 9% in the study, etc.	100% at the charging hub, 68% in the living room, 29% in the bedroom, etc.
Passenger	41% (the proportion of all passengers) adults, 22% kids, 19% the old, 15% pets, etc.	53% adults, 6% kids, 22% the old, 2% pets, etc.
Space layout	16% (the proportion of all conditions) on the carpet, 11% on the wood floor, 28% under the furniture, 22% near the wall, 7% on the humid floor, etc.	83% near the wall, 57% at the corner, 29% next to a shelf, 18% near a plant, etc.
Resource supply/received	4% (the proportion of all supplied or received resources) wools, 7% hairs, 7% leaves, 22% dirt, 14% data from the air conditioner, etc.	100% electrical power, 21% data from the centre hub, 40% data from smart home controllers, etc.
Notification	2.8% (the proportion of total notifications) notifications of cleaning progress, 2.1% notifications of volume left in the debris container, 1.7% of available battery volume, etc.	13.5% notifications of available battery volume, 4.1% notifications of software update, 26% notifications of environment monitoring, etc.
Interaction with peer products	3.3% (the proportion of total time of interactions) interactions with virtual walls, 14% interactions with smart home controller, 6.2% interactions with surveillance cameras, etc.	100% interactions with charging hub, 65% interactions with centre hub, 50% interactions with the smart home controllers, etc.
Service output	100% (the proportion of working time) 17–23 W suction power, 60% data exchange with smart phone, 15% notifications forwarded to users, etc.	35% notifications forwarded to users, 25% data exchange with centre hub, 18% data exchange with the smart home controller, etc.
Users' demography	43% (the proportion of users) are aged between 25 and 31, 61% are single, 38% live in apartment, etc.	43% are aged between 25 and 31, 61% are single, 38% live in apartment, etc.

(continued)

Table 4 (continued)

Contextual data	Mid-level cleaning mode	Charging mode
Users' mood	59% (the proportion of users) feel normal, 17% feel relaxed, 3% feel anxious, etc.	98% feel normal, 2% feel anxious, etc.
Using habits	64% (the proportion of users) use RVC during working time, 45% never use RVC to clean toilets, 57% use RVC twice a week, etc.	64% use RVC during working time, 45% never use RVC to clean toilets, 57% use RVC twice a week, etc.
User's preferences	68% (the proportion of users) prefer mid-level cleaning mode, 37% prefer RVC with the mopping function, 61% prefer the round shape RVC than the square shape, etc.	68% (the proportion of users) prefer mid-level cleaning mode, 37% prefer RVC with the mopping function, 61% prefer the round shape RVC than the square shape, etc.
Users' knowledge of products	37% (the proportion of users) have abundant knowledge about RVC, 21% have used more than four smart home products, 48% have never used RVC before, etc.	37% (the proportion of users) have abundant knowledge about RVC, 21% have used more than four smart home products, 48% have never used RVC before, etc.
Energy generation	16.63–18.31 W kinematic power, 21.74–23.86 W thermal power, 19.73–22.61 W suction power, etc.	11.22–13.07 W thermal power, 3.8–4.5 W electrical power
Command	43% (the proportion of all command) command of navigations, 22% command of kinematic control, 31% command of data exchange, etc.	61% command of data exchange, 32% command of environment monitoring, 4% command of power monitoring, etc.
Computation	21% (the proportion of all computation power) computation of navigation, 32% computation of statuses monitoring, 19% computation of kinematic control, etc.	28% computation of software updates, 26% computation of data transferring, 17% computation of data packing, etc.
Movement	62% (the proportion of all movements) linear movement at a uniform speed of 0.55 m/s, 10% 0.88 m/s^2 linear acceleration, 8% 1.57 rad/s rotational speed, etc.	Remain static
Energy consumption	65–80 W electrical power	13.5–15.5 W electrical power

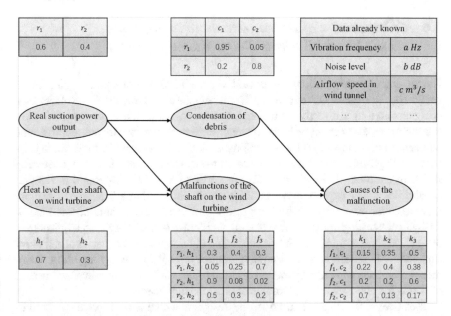

r_1	r_2
0.6	0.4

	c_1	c_2
r_1	0.95	0.05
r_2	0.2	0.8

Data already known	
Vibration frequency	$a\ Hz$
Noise level	$b\ dB$
Airflow speed in wind tunnel	$c\ m^3/s$
...	...

h_1	h_2
0.7	0.3

	f_1	f_2	f_3
r_1, h_1	0.3	0.4	0.3
r_1, h_2	0.05	0.25	0.7
r_2, h_1	0.9	0.08	0.02
r_2, h_2	0.5	0.3	0.2

	k_1	k_2	k_3
f_1, c_1	0.15	0.35	0.5
f_1, c_2	0.22	0.4	0.38
f_2, c_1	0.2	0.2	0.6
f_2, c_2	0.7	0.13	0.17

Fig. 4 Example of the diagnosis for robot vacuum cleaner

of each potential cause). Note that this is a simplified case of Bayesian Network for illustrative purposes only; a practical diagnosis requires a far greater level of sophistication.

In addition, DT provides a revolutionary predictive capability for an RVC's status based on mathematical models and virtual simulations. For example, the roller shaft is an RVC's most consumable parts. Based on constant monitoring of performance over time, the DT can predict the shaft's fatigue or failure rate and remind the user to replace worn parts in advance. The mechanism by which this prediction is made is as follows. Assume that a crack of $c\ mm$ is detected on the shaft, and DT will predict when the crack will grow up to $b\ mm$. The prediction can be based on the Paris Law $da/dN = C(\Delta K)^m$ [21]. In the formula, da is the change of crack length and dN is the experienced load cycle. Different shaft behaviours present varying load cycles obtained from the PE and TDC; C and m are material coefficients obtained from online resources; and ΔK is the range of stress intensity recorded by the PE. Based on shaft stress historical recordings, shaft load cycles of each working period, and material resources, the DT can forecast the moment when the crack grows up to $b\ mm$. However, environmental factors (humidity, acid-base, users' activities) also affect crack growth, but these factors are not contained in the formula. In such conditions, DT can estimate crack growth based on statistical probabilities. Here, PE must model all environmental factors and evaluate their influences on crack propagation. For example, higher air humidity levels may accelerate corrosion, further increasing crack growth. The DT constantly improves based on this feedback loop between the PE and VE, i.e., the more crack growth data recorded, the more accurate future predictions will become.

5.4   Enhancement of Human-Machine Interactions via the Virtual Entity of Robot Vacuum Cleaner

Human-machine interactions conducted via the RVC VE offer many benefits to designers. For example, during RVC manufacture and maintenance processes, designers no longer directly control and handle production machines such as lathes, robotic arms, and conveyor belts. Instead, designers can investigate, control, and manoeuvre those machines by controlling their VEs in dedicated panels. Such indirect manoeuvring overcomes constraints of physical distance and users' manoeuvring skills. In addition, designers may also use VEs to test the suction and navigation capabilities under extreme environments. Such tests enable a designer to explore new potential uses while prohibiting safety problems. An example of this kind of designer-led testing could involve placing an RVC in a laboratory with adjustable − 30–185 °C temperature, which is inaccessible to human without special protections. Enabled by DT, designers can remotely control the RVC status by inputting movement commands and output power. This kind of tests help designers evaluate usages of RVCs for polar research or sample collections in volcanos.

Meanwhile, the RVC VE provides the designers with micro-observations of individual components' performance. Changing parameters such as coordinates, rotating angle, and airflow speed enable designers' precise control over RVC operation. Furthermore, remote operation using the VE of a customer's RVC enables rapid resolution of malfunctions. For example, if a customer's RVC produces abnormal noises when cleaning, a customer may notify the problem and seeks for designers' troubleshoots remotely. Here designers use the RVC's VE to check the status of relevant components. From observations of the virtual model, designers may notice the undamped vibration of the wind turbine. Based on this phenomenon, designers then check the virtual model of buffers and find a fractured one. The outcome and maintenance advice will be sent back to the customer.

In addition to benefitting designers', human-machine interactions via RVC VEs benefit customers profoundly. One example involves the ease with which users can troubleshoot operational issues or replace worn components. Animations on the VE may provide customers with highly detailed instructions and even alert customers to future problems. Furthermore, the additional transparency DTs provide of the manufacturing process strengthens the trust between users and products. For example, users may investigate the quality checks or how a product is made, which proves to them how reliable one product is over another. Customers may also participate in the design process, which improves product quality and strengthens loyalty between users and brands. Moreover, DT promotes customers' engagement in RVC design. Some customers may want to customize design configurations of their RVCs, such as geometry design, aesthetic design, and optional functions. To forward their ideas, customers can choose those configurations in VE, which will automatically build virtual prototypes and forward to designers. Meanwhile, designers can polish those prototypes and send demos of virtual products back to customers via VE as well.

6 Conclusion

This chapter envisions how DT can support the data-driven engineering design. Data-driven engineering design has three crucial challenges: the amount of data, the quality of virtual models, and the lead time. In corresponding, DT has three essential features that resolve the three challenges: (1) DT comprehensively and dynamically collects data from abundant sources; (2) DT significantly improves the similarity of virtual models; and (3) DT achieves the synchronization between PE and VE. DT provides three services to support the data-driven engineering design, including (1) real-time monitoring and data collections in the physical world; (2) identifications, diagnoses, and predictions of product statuses; (3) enhancement of human-machine interactions via VE for designers and customers. As an emerging technology, DT is attracting more attention from many scholars and will play an important role in the future era of smart manufacturing.

References

1. Pahl, G., & Beitz, W. (2013). *Engineering design: A systematic approach*. Springer Science & Business Media.
2. Todd, R. H., & Magleby, S. P. (2004). Evaluation and rewards for faculty involved in engineering design education. *International Journal of Engineering Education, 20*(3), 333–340.
3. Chang, T. C., & Wysk, R. A. (1997). *Computer-aided manufacturing*. Prentice Hall PTR.
4. FÜrst, K., & Schmidt, T. (2001). Turbulent markets need flexible supply chain communication. *Production Planning & Control, 12*(5), 525–533.
5. Glaessgen, E., & Stargel, D. (2012,). The digital twin paradigm for future NASA and US Air Force vehicles. In *53rd AIAA/ASME/ASCE/AHS/ASC Structures, Structural Dynamics and Materials Conference 20th AIAA/ASME/AHS Adaptive Structures Conference 14th AIAA* (p. 1818).
6. Tao, F., et al. (2019). Five-dimension digital twin model and its ten applications. *Computer Integrated Manufacturing Systems, 25*(1), 1–8.
7. Qi, Q., & Tao, F. (2018). Digital twin and big data towards smart manufacturing and industry 4.0: 360 degree comparison. *IEEE Access, 6*, 3585–3593.
8. Boschert, S., & Rosen, R. (2016). Digital twin—the simulation aspect. In *Mechatronic futures* (pp. 59–74). Springer, Cham.
9. Yun, S., Park, J. H., & Kim, W. T. (2017, July). Data-centric middleware based digital twin platform for dependable cyber-physical systems. In *2017 Ninth International Conference on Ubiquitous and Future Networks (ICUFN)* (pp. 922–926). IEEE
10. Tomizawa, M., Yamabayashi, Y., Sato, I., & Kataoka, T. (1994). Nonlinear influence on PM-AM conversion measurement of group velocity dispersion in optical fibres. *Electronics Letters, 30*(17), 1434–1435.
11. Brown, A. L., Yoon, S. S., & Jepsen, R. A. (2008). Phenomenon identification and ranking exercise and a review of large-scale spray modeling technology. In *ASME 2008 Heat Transfer Summer Conference collocated with the Fluids Engineering, Energy Sustainability, and 3rd Energy Nanotechnology Conferences* (pp. 569–579). American Society of Mechanical Engineers.
12. Suh, S. H., Seo, Y., Lee, S. M., Choi, T. H., Jeong, G. S., & Kim, D. Y. (2003). Modelling and implementation of internet-based virtual machine tools. *The International Journal of Advanced Manufacturing Technology, 21*(7), 516–522.

13. Núñez, H., Angulo, C., & Català, A. (2002). Rule extraction from support vector machines. In *Esann* (pp. 107–112).
14. Sarawagi, S., Thomas, S., & Agrawal, R. (1998). *Integrating association rule mining with relational database systems: Alternatives and implications* (Vol. 27, No. 2, pp. 343–354). ACM.
15. Cattell, R. (2011). Scalable SQL and NoSQL data stores. *Acm Sigmod Record, 39*(4), 12–27.
16. Bar-Shalom, Y., Willett, P. K., & Tian, X. (2011). *Tracking and data fusion.* CT, USA: YBS Publishing.
17. Bailey, T. C., & Gatrell, A. C. (1995). *Interactive spatial data analysis* (Vol. 413). Essex: Longman Scientific & Technical.
18. Qi, Q., Tao, F., Hu, T., Anwer, N., Liu, A., Wei, Y., Wang, L., & Nee, A. Y. C. (2019). Enabling technologies and tools for digital twin. *Journal of Manufacturing Systems.* (Duan, L. M., Lukin, M. D., Cirac, J. I., & Zoller, P. (2001). Long-distance quantum communication with atomic ensembles and linear optics. *Nature, 414*(6862), 413).
19. Judd, G., & Steenkiste, P. (2003). Providing contextual information to pervasive computing applications. In *Proceedings of the First IEEE International Conference on Pervasive Computing and Communications, 2003.(PerCom 2003).* (pp. 133–142). IEEE.
20. Chickering, D. M. (2002). Learning equivalence classes of Bayesian-network structures. *Journal of machine learning research, 2*(Feb), 445–498.
21. Paris, P. C., & Erdogan, F. (1963). A Critical Analysis of Crack Propagation Laws. *Journal of Basic Engineering, 85,* 528–533.

Chapter 9
Enabling Technologies of Data-Driven Engineering Design

Abstract This chapter presents a collection of specific technologies that can collectively enable data-driven engineering design. The enabling technologies include sensors, Internet and mobile Internet, Internet of Things, cloud computing, edge computing, blockchain, machine learning, artificial intelligence, big data analytics, virtual reality and augmented reality, digital twin, and so forth. These enabling technologies are presented in correspondence to relevant data operations such as data collection, transmission, computation, analysis, security, and cyber-physical integration. The enabling technologies can be integrated, either vertically or horizontally, towards an integrated system of systems. Multiple specific applications of these enabling technologies are presented as illustrative examples.

Keywords Sensing technology · Cloud computing · Edge computing · Internet · Internet of Things · Machine learning · Artificial intelligence · Blockchain · Digital twin · Virtual reality · Wearable device

1 Introduction

The practical implementation of data-driven engineering design depends on a variety of specific enabling technologies for the corresponding data operations. Data-driven design is by nature an interdisciplinary, if not transdisciplinary, endeavour that involves numerous stakeholders, technologies, and tools from different disciplines (e.g., design science, data science, computer science, and cognitive science). Depending on the data operations involved in a data-driven design solution, a variety of enabling technologies should be synthesized in different fashions. Identical to the roles of oils in driving the traditional enabling technologies in the machine age, data plays critical roles in driving the modern enabling technologies.

In different eras, the success of engineering design hinges on different enabling technologies. The enabling technologies of engineering design used to include computer-aided design (for concept generation and evaluation), Internet (for understanding customer voice), simulation (for design analysis and optimization), etc. Since 2010, a number of new enabling technologies have emerged. In particular, the major economies (e.g., the USA, European Union, China, and Japan) have put

© The Author(s), under exclusive license to Springer Nature Switzerland AG 2022   173
A. Liu et al., *Data-Driven Engineering Design*,
https://doi.org/10.1007/978-3-030-88181-8_9

forward different national initiatives such as Industrial Internet, Industry 4.0, Made in China 2025, Society 5.0, and so forth. Despite their natural differences with respect to objective, context, and priority, these initiatives more or less share some common value propositions such as the connection of intelligent machines to exchange information, the immersive integration between physical and digital worlds, the incorporation of information and communication technologies (ICTs) into the design and manufacturing processes, and the autonomous decision-making towards higher efficiency and productivity. These initiatives in turn promoted the emergence of many new enabling technologies that can benefit data-driven engineering design.

In the previous chapters, the focus lies in the integration between data operations and design operations from the scientific, theoretical, and methodological perspectives. As the finale of this book, this chapter presents a collection of enabling technologies that can serve as practical means to realize different data operations. The enabling technologies of data-driven design include but are not limited to sensor, Internet, mobile Internet, Internet of Things, blockchain, cloud computing, edge computing, machine learning, artificial intelligence, big data analytics, digital twin, virtual reality, wearable device, etc. For each category of enabling technologies, discussions will surround their historical evolvement, technological advantages and disadvantages, practical applications in different industries, and specific applicability to the particular paradigm of data-driven engineering design.

The rest of this chapter is organized as follows in correspondence to different stages of a typical data lifecycle. Section 2 elaborates the enabling technologies for the collection, transmission, and connection of design data. Section 3 explains the enabling technologies for the storage, computation, security, and privacy of design data. Sections 4 and 5 present the enabling technologies for the analysis and cyber-physical integration of design data, respectively. Section 6 illustrates the systematic synthesis of various enabling technologies from the perspective of System of Systems (SoS). Section 7 presents some illustrative examples of adopting specific enabling technologies to address practical design problems. Section 8 draws conclusions.

2 Enabling Technologies of Data Collection and Transmission

2.1 Sensing Technologies

A sensor functions to detect changes of a target event or attribute within a particular environment. Various sensors constitute the technological foundation of data acquisition in the physical world. Since typical sensors cannot directly process data, data acquired by sensors are subsequently transmitted to computer processors for further processing. Made possible by sensors, human operators can be freed from manual data acquisition, which is both time-consuming and cost-ineffective. Such a freedom makes it possible to collect data continuously without human-related interruptions.

In practice, sensors tend to exist as an integral component of an integrated system, as opposed to an independent artefact. Take the design of smart products, for example, the commonly used sensors include temperature sensor, pressure sensor, infrared sensor, touch sensor, humidity sensor, flow sensor, smoke sensor, proximity sensor, light sensor, gyroscope, accelerometer, ultrasonic sensor, etc.

The quality of a sensor can be assessed by a variety of metrics such as sensitivity (i.e., to what extent the sensor can detect minor changes), accuracy (i.e., to what extent the acquired data can truthfully reflect physical realities), robustness (i.e., to what extent the sensor can perform regularly against various uncertainties), etc. Ideally, a sensor is expected to detect changes in an environment without imposing extra impacts to the environment. For example, the heat produced by a temperature sensor should impose minimum impacts on the environment temperature.

Practical applications of sensors can be found ubiquitously in the modern society. In particular, sensors play irreplaceable roles in the design of smart products, autonomous vehicles, cyber-physical systems, industrial product-service systems, cyber-physical production systems, etc. In the scenario of data-driven design, sensors can be employed to collect data concerning customers, products, and environment in the physical world. Some sensors (e.g., touch sensor and pressure sensor) can be utilized to recognize customer behaviours and reflect customer-product interactions. Such sensors are commonly deployed in the design of mobile devices and wearable devices that are in close proximity to the target customers. Some sensors can be used to detect changes and interact with an external environment. For example, self-driving cars rely on camera, radar, and lidar sensors to detect obstacles in a surrounding environment [1]. Based on such data, self-driving cars can be made more autonomous. Besides, sensors can be used to monitor various environments (e.g., house, office, farm, transportation, factory, etc.) with respect to different combinations of environmental changes concerning weather, temperature, humidity, wind, lighting, and so forth [2]. Lastly, sensors can be used to monitor the state of a product in terms of its function, behaviour, and performance. For example, a commercial airplane is typically equipped with thousands of sensors to monitor different facets of the airplane in real time in the interest of product lifecycle management (PLM).

2.2 Internet and Mobile Internet

The dynamic flow of design data relies on a network of interconnected terminals. Internet is essentially a network of interconnected computers that can communicate information and facilitate collaboration. Technically, the Internet as we know today is an enormous 'system of systems' that includes numerous networks of various scales and different natures. In some sense, it can be argued that, without the Internet, the majority of enabling technologies of data-driven design would not have been possible. Since its advent in the 1990s, the Internet has revolutionized numerous industries, leading to countless new business models such as e-commerce, online entertainment, social media, social network, virtual collaboration, etc. It is by no means overstating to claim that the Internet constitutes one of the most important fundamental infrastructures in the modern society.

The Internet also constitutes an integral element of data-driven design. The Internet by itself accommodates countless data, the majority of which is massively available and freely accessible, which can enable designers to understand customer preference, customer journey, competitor offering, market dynamics, product benchmarking, etc. Furthermore, the Internet makes it possible for a large cohort of globally distributed stakeholders (e.g., customer, designer, regulator, manufacturer, supplier, and service provider) to collaborate virtually on the same design project across national, institutional, and physical boundaries.

Due to the increasing popularity of mobile devices (e.g., smartphone and tablet computer), the traditional computer-based Internet is increasingly outperformed, if not replaced in some sense, by the mobile Internet [3]. The mobile Internet originally emerged as an extension and hence a subset of the conventional Internet. Specifically, mobile devices were initially employed as a more portable and convenient means to access information that is readily available on the conventional Internet. Triggered by the growing popularity of mobile devices, however, more and more dedicated software, content, and business models are being developed solely for the mobile Internet. As a consequence, today's mobile Internet is evolving growingly independent from the traditional Internet. In particular, the advancement of telecommunication technologies, represented by 5G and powerful processers, continues to push the physical boundaries of data transmission and computation for the mobile Internet.

In comparison with data obtained from the conventional Internet, data obtained from the mobile Internet affords unique values for product and service design. Different from computers (i.e., the basic nodes of the conventional Internet) that are mostly used in certain spaces (e.g., office and home) at certain times, customers are more used to carrying mobile devices all the time, leading to a more continuous and broader spectrum of data that can reflect the holistic picture of customer habit, preference, expectation, demographics, journey, and differentiation. Moreover, data generated by mobile devices can reflect the (nearly) real-time product states, which is important for those design scenarios that depend on real-time data to adjust product configurations in a timely manner. Data generated by a mobile device can accurately reflect the contexts in which the device is used in terms of location, weather, surrounding environment, etc.

2.3 Internet of Things (IoT) and Industrial Internet of Things (IIoT)

Internet of Things (IoT) refers to a network of physical 'things' that are interconnected to exchange data/information with each other. Different from the Internet and mobile Internet that rely heavily on the person behind a computer or smartphone, IoT empowers intelligent machines to collect, transmit, and exchange data in a more autonomous fashion. In specific to product development, one of the most well-known IoT scenarios is smart home, where various smart appliances and devices are connected to exchange information towards synchronized actions.

By connecting multiple physical things into a grand network, IoT can result in enormous values. Firstly, IoT makes it possible for interconnected physical product/devices to exchange data, share information, and benefit from each other's functionalities. For example, even though a typical air conditioner is not equipped with the capability of weather forecast, it can still acquire the relevant weather information from other more capable devices such as a smartphone or a smart speaker. Secondly, IoT enables a physical product/device to dynamically adapt the behaviours and optimize the performance in light of information acquired from the IoT network. For instance, based on the weather information collected from a smart speaker, a smart air conditioner can dynamically adjust the room temperature and optimize the energy efficiency. Lastly, IoT enables a consortium of products that are connected to the same network to collaboratively perform tasks that cannot be accomplished by individual products independently. For example, by connecting adjacent self-driving cars and autonomous drones, they can share traffic information with each other to achieve desirable herd behaviours.

Practical applications of IoT can be found in numerous industries [4, 5]. In transportation, various IoT solutions are developed to trace vehicle health, monitor traffic, reduce congestion, optimize route planning, and track road conditions. In agriculture, IoT constitutes the technological backbone of smart farms, where the soil condition is continuously monitored, irrigation is dynamically controlled, and diseases are identified at the earliest opportunity [6]. In supply chain, IoT is employed to monitor shipment, reinforce inventory management, and realize predictive maintenance [7]. In healthcare, IoT has been incorporated into many medical devices to monitor the real-time location, operation, and health of not only the device but also the patient who interacts with the device [8]. In particular, IoT establishes the technological foundation of various wearable devices.

Industrial Internet of Things (IIoT) can be regarded as a particular kind of IoT network that is tailored into industrial scenarios, in which the connected 'things' refer specifically to industrial equipment, machinery, process, and device. Practical applications of IIoT can be found in manufacturing, logistics, transportation, energy, smart city, and so forth [9]. A typical scenario for the application of IIoT is smart factory [10, 11], where various machines are connected, monitored, and controlled in real time to achieve higher productivity, reliability, efficiency, and sustainability. Another typical scenario is agriculture, where traditional farms are transformed towards 'smart farms' by monitoring the livestock health and location, tracking farm conditions, controlling farm irrigation, managing livestock feeding, regulating agriculture vehicles, etc.

Data security is a major concern for both IoT and IIoT. Once a set of physical terminals are interconnected towards a network, successful attacks on an individual node could potentially jeopardize the security of the whole network. Unlike cyberattacks on the Internet that can only affect the digital world, attacks on the IoT and IIoT networks could potentially influence many facets of the physical world. Furthermore, data privacy is an equally significant concern. Such a concern is especially true for the scenario of smart home and smart hospital, the data from which can reflect highly sensitive customer/patient behaviours, habits, and preferences. Therefore, if any IoT or IIoT data would be included in a data-driven design solution, it is necessary to develop additional mechanisms to protect data security and privacy.

Both IoT and IIoT are highly applicable to data-driven engineering design. IoT paves the way for transforming a regular product into a smart product that can sense the surrounding environment, interact with other smart products, and request for value-adding services. After a ubiquitous IoT network of interconnected products is established, it becomes possible for peer products to collaborate with each other in order to achieve functionalities that are otherwise impossible for a single product. Moreover, both IoT and IIoT are key enabling technologies for product lifecycle management (PLM), in which a key task is to obtain relevant data that can reflect a product's states throughout different stages of its lifecycle. Such data is especially important for industrial product-service systems (IPSS).

3 Enabling Technologies of Data Storage and Computation

3.1 Cloud Computing

Cloud computing refers to the on-demand offering of computing resources with respect to data storage, computation, integration, analysis, visualization, etc. Different from the traditional paradigm of exclusive, local, and centralized computing, cloud computing makes it possible for enterprises to access computing resources as per need in the format of computing services [12, 13]. Cloud computing is a fundamental infrastructure for most of the data-driven endeavours. Cloud computing can be classified into public cloud, private cloud, and hybrid cloud. Public cloud is readily accessible, through subscriptions, to different kinds of users (corporate as well as individual users) based on the public Internet. In contrast, private cloud is intended to accommodate computing tasks associated with a single enterprise, hence private to the general public. Hybrid cloud, as the name suggests, proportionally distributes the functionalities, resources, and capabilities of cloud computing into both public and private clouds. As the security level of public cloud continues to improve, more and more enterprises are embracing the more cost-effective public cloud.

A distinguishing feature of cloud computing is a service-oriented business model [14, 15]. Instead of investing heavily to build, maintain, and upgrade dedicated computing resources, through cloud computing, an enterprise can still acquire the necessary computing resources from third-party service providers such as Microsoft Azure, Amazon Web Services (AWS), Google Cloud, Alibaba Cloud, Tencent Cloud, etc. In addition to data storage, these service providers also supply other value-adding services such as artificial intelligence, data visualization, data analytics, etc. Hence, cloud computing can potentially serve as a one-stop platform for manufacturers to accommodate the comprehensive digitalization of their design operations, processes, and projects.

Cloud computing is featured with some notable advantages. Firstly, benefited from the economics of scale, cloud computing can significantly reduce the cost of data storage. Cost-effectiveness is especially important for small-scale design projects run by small- and medium-sized enterprises. Secondly, cloud computing can lead to greater flexibility and scalability, as the amount of computation resources required for a design task can be dynamically adjusted in alignment with customer needs. Thirdly, cloud computing can enhance productivity for those design scenarios that involve globally distributed stakeholders, who can analyse the same dataset stored in the cloud simultaneously. Lastly, cloud computing makes it possible to reduce data silos by prompting inter-organizational collaborations in the cloud.

Practical applications of cloud computing can be found in many industries such as education, entertainment, manufacturing, supply chain management, etc. In particular, the integration between cloud computing and IoT resulted in the paradigm of cloud manufacturing [16–18] through which globally distributed manufacturing resources can be accessed and allocated in the format of on-demand manufacturing services (e.g., design, customization, rapid prototyping, production, maintenance, testing, etc.).

Cloud computing plays critical roles in data-driven engineering design. At the corporate level, it enables small- and medium-sized enterprises to develop new value-adding services (e.g., data-driven services such as predictive maintenance and prognosis [19]) that can counterbalance the cost of cloud computing. At the product level, cloud computing is especially applicable to smart products (e.g., smart appliances in a smart home) that are equipped with specialized functions and hence constrained local computing resources. With respect to project management, cloud computing paves the way for more effective collaborative engineering in product development. It should be made clear that cloud computing is not a universal formular. Since it involves data transmission to and from the cloud, it introduces new issues such as data ownership, privacy, and security. Besides, the reliability of cloud computing becomes questionable in the scenario of abruptly and unexpectedly surging requests for computing resources at the same time.

3.2 Edge Computing

Edge computing represents a distributed paradigm of computing that enables a product/device to store, query, and compute data near, if not inside, the product/device [20–23]. Compared to cloud computing that still follows a centralized paradigm (i.e., the cloud serves as a central node), edge computing is designed to thrust the data storage, query, and computation from the central cloud towards the local edges.

Compared to centralized computing, edge computing has some notable advantages concerning privacy protection, communication efficiency, system reliability, etc.

Above all, edge computing is more compliant with the growingly rigorous regulation on data privacy, because it can effectively prevent raw data from being transmitted to a central server without customer consent. On the other hand, edge computing introduces new challenges to the protection of data security [24]. New encryption approaches are needed to cope with direct cyberattacks against individual edges. Secondly, a peer-to-peer network that is composed of interconnected devices with analogous edge computing capabilities is by architecture more reliable than a star network whose central node tends to expose a single point of failure. Thirdly, since edge computing enables a product/device to store and compute data locally in place of transmitting raw data to a central server, it may reduce some communication cost. Lastly, edge computing can make a system more responsive to uncertainties as well as complexities through faster data query and on-site computation.

Practical applications of edge computing can be found in scenarios that have high requirements on real-time computation. In transportation, edge computing can be used to connect, regulate, and command autonomous vehicles, drones, and traffic infrastructure. In healthcare, edge computing can be used to power wearable devices that are designed to monitor emergency situations that require immediate medical attention and care. In manufacturing, edge computing can be used to monitor equipment health, control machine operations, and prescribe maintenance services on site based on real-time data [25]. In smart city, edge computing plays important roles in remotely monitoring key infrastructure such as smart grid and gas plant [26]. In all these applications, it is desirable, if not required, that data should be computed at the earliest opportunity to avoid immediate risks of operability, safety, and security.

It should be made clear that cloud computing and edge computing are not mutually exclusive technologies. They can be engaged to play complementary roles in data-driven engineering design. Cloud computing is more applicable to the design scenarios (e.g., model building and iterative optimization) that involve time-insensitive data, high tolerance on latency, reliable bandwidth, and high requirement on computation. In contrast, edge computing is more suitable to the design scenarios (e.g., quality inspection and autonomous decision-making) that involve time-sensitive data, context-dependent uncertainty, real-time analysis, unreliable bandwidth, and relatively low requirement on computation.

Some advantages of edge computing make it especially applicable to the design scenarios that are featured with high requirements on customization, autonomy, and intelligence. For the consumer market, edge computing is applicable to the design of various smart products, which are characterized by high degrees of personalization, reconfigurability, and autonomy. Besides, edge computing can be employed to develop smart product-service systems that require real-time data processing. For the industrial market, edge computing is a key enabling technology for the cyber-physical systems (CPS) and cyber-physical production systems (CPSS) that have high requirements on reliability and computation speed [27].

3.3 Blockchain

Blockchain refers to a digital ledger that consists of numerous interdependent records, namely, blocks, to document transactions across numerous nodes in a network [28, 29]. Blockchain makes it possible for a large network of distributed stakeholders to collaboratively store, manage, and verify data in a greatly transparent and cryptographic fashion. Blockchain enables a network of distributed stakeholders, who needlessly trust each other, to reach a decentralized consensus, which cannot be manipulated by any individual stakeholder or central node. Depending on the permission of data access, blockchains can be classified into public blockchain (i.e., data is accessible to any node in the network), private blockchain (i.e., data can only be accessed by permitted nodes), and hybrid blockchains (i.e., some data is public whereas some other data is private).

One of the most prominent features of blockchain is immutability (i.e., modification resistance). The unique architecture of blockchain ensures that it is difficult, if not practically impossible, to deliberately modify a record documented in a block without affecting other blocks. The intra-block communication and validation of data is enabled by a peer-to-peer network. Coupled with the feature of immutability, a public blockchain is also characterized by high levels of transparency and openness with respect to data access, inquiry, and transfer. Finally, blockchain is a thoroughly decentralized solution that can potentially empower a large network of globally distributed nodes (and the stakeholders behind) to negotiate, maintain, and update a decentralized consensus [30], which coincides with one of the key pursuits of Industry 4.0 [31].

Furthermore, blockchain makes it possible to develop and deploy smart contracts, which refer to digital contracts that can be triggered, enforced, and documented directly based on computational logic with little human involvement [32]. Smart contract can significantly reduce the reliance of traditional contracts on trusted mediators that tend to yield high cost and prolong contract processing. Smart contract is applicable to product design in many ways. As explained in Chap. 6, smart products demand customized services that can be triggered automatically. Hence, smart contract is useful for reducing the engagement of third parties in an iterative customization process [30]. As data is growingly regarded as a new asset, smart contracts are needed to facilitate the convenient transaction of digital assets without intermediators.

Practical applications of blockchain can be found in many industries. Majority of blockchain applications can be found in the financial industry, where the technology is used to issue cryptocurrencies, document financial transactions, exchange digital assets, and so forth [33]. In supply chain management, blockchain can be used to trace the provenance of supplies and track the transportation/delivery in mining, food, and shipping [34]. In manufacturing, blockchain can be used to enhance quality assurance, promote product customization, identify counterfeits, and so forth [35]. In e-commerce, blockchain can be used to ensure that online customer reviews cannot be altered or deleted. In healthcare, blockchain can be used to document, manage, and share sensitive medical data [36]. For IoT and IIoT, blockchain can be used to enhance the cybersecurity of IoT devices in the context of smart home and smart city [37]. In public service, blockchain can be used by governments to manage citizen identity, archive public documents, store private records, and enhancing voting integrity.

Blockchain can be incorporated into data-driven design to enhance data transparency and mutual trust. It can be used to protect the intellectual property associated with a design concept, prototype, or patent that can be reflected by data. It can enable multiple stakeholders, who unnecessarily trust each other, to jointly negotiate, document, and manage a decentralized consensus concerning product design, which is a longstanding challenge of collaborative engineering [38]. It can enable relevant stakeholders (e.g., manufacturer, customer, and regulator) to trace the provenance of raw materials, which is especially needed for certain design scenarios that involve a high level of customization and personalization.

4   How to Analyse and Understand Data

4.1   Machine Learning

Machine learning makes it possible for an intelligent product, agent, or system to perform new tasks through self-learning instead of pre-defined programming. Machine learning focuses on learning from past events to predict future events. In general, various machine learning approaches can be classified into three main categories: supervised learning, unsupervised learning, and reinforcement learning.

Supervised learning is intended to transform a given input to an unknown output in light of the pre-defined and prelabeled input-output mappings. The model training of supervised learning is conducted based on labelled dataset. Commonly used algorithms of supervised learning include support vector machines, linear regression, logistic regression, probabilistic classifier, decision tree learning, k-nearest neighbour algorithm, etc.

In contrast to supervised learning, unsupervised learning is intended to cope with unlabeled data. Clustering and anomaly detection are two common approaches of unsupervised learning. Clustering functions to divide a number of items into different clusters, for which, items grouped into the same cluster are more similar to each other than with items in other clusters [39]. Specific clustering algorithms include connectivity-based clustering (i.e., the clusters are determined based on connections among items), centroid-based clustering, distribution-based clustering, density-based clustering, grid-based clustering, etc. Unlike clustering that focuses on the item similarity, anomaly detection (or outliner detection) functions to reveal items that are significantly different from other items in a dataset [40].

Reinforcement learning investigates how intelligent machines should act, in a foreign environment that is full of uncertainties, in order to maximize rewards and minimize punishments [41, 42]. Reinforcement learning is especially useful for enhancing the intelligence of various smart products. It can be used to coordinate actions of a consortium of interconnected products (e.g., various smart appliances situated in a shared smart home environment) with respect to their exchange, collaboration, and competition for rewards. Reinforcement learning is widely used in the design of video games. In transportation, reinforcement learning can be used to enhance the intelligence of autonomous vehicles with respect to path planning, energy optimization, and risk identification [43]. In smart factories, reinforcement learning can be used to train industrial robots by rewarding desirable robotic behaviours (e.g., precise or safe operations) as well as penalizing undesirable behaviours (e.g., inaccurate or dangerous operations) [44].

Practical applications of machine learning can be found in many industries. In manufacturing, it can be used to identify various anomalies (e.g., quality defect, inconsistency in a manufacturing process, safety risk in human-robot collaboration, etc.), automatically optimize a design concept against known and even unknown design constraints, analyse images for quality inspection, and enable preventative maintenance of production equipment [45, 46]. In healthcare, machine learning can be used to predict a patient's health risk, support pathologists to arrive at more informed diagnosis based on medical imaging data, evaluate the benefit/risk of different treatments, conduct medical research that is too complex or time-consuming for human researchers, develop new drugs, match patients with relevant doctors, and so forth [47]. In finance, machine learning can be used to detect frauds and money laundering, moderate dynamic pricing, and recommend financial news and services to investors. In specific to product development, recommendation system can be used to recommend products to target customers, anomaly detection can be used for quality assurance, clustering can be used to group identical customers, reinforcement learning can be used to generate optimized structures against known design constraints, etc.

As a subset of machine learning, deep learning is designed to simulate the working mechanism of human brains through various kinds of artificial neural networks such as deep neural network, recurrent neural network, and convolutional neural network [48]. Practical applications of deep learning include speech recognition, behaviour recognition, image recognition, natural language processing, medical diagnosis, mobile advertising, fraud detection, etc. Some of these applications are directly relevant to product design. For example, the intelligent recognition of human speech, motion, and image can enhance user-product interactions. Deep learning has notably high requirements on computing resources (e.g., high-performance GPU) as well as the abundance of labelled data. Machine learning has greater flexibility than deep learning in terms of the combination of models, classifiers, and features. Deep learning is therefore more suitable for the design scenarios that demand high prediction accuracy at the cost of high computation.

4.2   Artificial Intelligence

Artificial intelligence (AI) refers to the intelligence exhibited by artefacts as opposed to biological systems. According to the definition of 'artefact' by Herbert Simon in his pioneering work 'The Sciences of The Artificial' [49], artefacts are 'man-made things' that are both purposeful and functional. From the perspective of engineering design, AI can be regarded as a particular kind of engineered capability, which is purposefully assigned by designers to an artefact, in order to mimic the corresponding capability of human cognition, especially with regard to learning, problem-solving, and decision-making. Since its initial appearance in the 1950s, the research and application of AI have experienced multiple upward and downward waves [50]. AI is born to be an interdisciplinary subject matter, the scientific foundation of which can be traced back to computer science, statistics, cognitive science, mathematics, etc. AI has been employed to address typical problems such as natural language processing, reasoning, problem-solving, knowledge extraction and representation, etc.

Strictly speaking, some studies of machine learning can be regarded as a subset of artificial intelligence. In fact, there exist great overlapping among artificial intelligence, machine learning, and data science. Here the two notions (i.e., AI and machine learning) are discussed separately because they play different roles in data-driven design. Specifically, machine learning serves to engage machines to make predictions based on the analysis of data, and artificial intelligence functions to enhance the intelligence of machines to mimic human cognition. The subtle distinction is based on an underlying assumption that an intelligent design machine may eventually develop new cognitive patterns and abilities that are notably different from human cognition (hence beyond the strict scope of AI).

As artificial intelligence continues to advance, some new challenges are arising. It is increasingly difficult, if not impossible, for human users to explicitly understand the 'thinking', 'reasoning', and 'decision-making' process of intelligent machines [51]. As a result, it is challenging to clearly differentiate a 'bad' machine decision due to the lack of intelligence from a 'good' machine decision that cannot be interpreted by human. The design of smart products with AI capabilities involves a number of ethical issues concerning the interaction between AI and human, for example, to what extent AI should be hold accountable for the basic principles of humanity, how to deal with the unemployment of human workers caused by the large-scale deployment of AI, how to avoid the biased opinions of AI that can be triggered by the bias of data, etc. Some of these ethical issues are made even more complicated by the 'black-box' nature of AI. On top of the traditional metrics such as quality, reliability, robustness, efficiency, and accuracy, the development of AI-enhanced product and service must also consider new metrics such as security, responsibility, explainability, ethics, transparency, and privacy.

Practical applications of artificial intelligence can be found in many industries. In healthcare, AI is used to evaluate health risk, diagnose disease based on medical imaging data, analyse drug-drug interactions, develop new drugs, etc. In e-commerce,

AI is employed to make recommendations to target customers, personalize the shopping experience, provide responsive customer services, etc. In manufacturing, AI is playing important roles in human-robot collaboration, quality assurance, product life-cycle management, predictive maintenance, etc. In transportation, AI is proven useful for driving autonomous vehicles, managing traffic, enhancing safety and security, etc. In science, AI is employed to process complex dataset, discover new chemical compounds, explore uncharted outer space, and discover new biological structures.

Artificial intelligence can benefit data-driven design in multiple ways. AI can enhance the intelligence of not only smart products but also smart factories. In particular, AI is becoming an integral component of various smart products, autonomous systems, and wearable devices. Furthermore, the advancement of artificial general intelligence (AGI) is expected to result in intelligent design 'assistants', which (or more appropriately 'who') can assist human designers to gather, interpret, and visualize data. Identical to the roles of virtual assistants in smart products (e.g., Siri for iPhones and Alexa for smart speakers), virtual design assistant is an important technological means to engage machines in a design process. Some efforts have been devoted to developing intelligent agents to facilitate the extraction, representation, management, and reusage of design knowledge [52]. Finally, AI can be used to reinforce product-service integration for product-service systems (PSS) through AI-enhanced customer service and digital assistant.

4.3 Big Data Analytics

Big data refers to datasets that are too large to be analysed primarily by human operators by means of traditional data methods, techniques, and tools. Big data is characterized by a set of 'V's, namely, volume (i.e., the quantity of data), velocity (i.e., the speed and pace at which data is being generated), variety (i.e., to what extent data varies from each other), value (i.e., the importance or usefulness of data to business and decision-making), veracity (i.e., the quality and reliability of data), etc. The high complexity associated with big data imposes great challenges to a number of data operations such as data acquisition, transmission, processing, analysis, visualization, etc. Due to the growing cyber-physical integration, data is being accumulated from both physical and digital worlds, which further escalates the difficulty of big data analytics. It is worthwhile to iterate that the interest level in 'big data' had been higher than that in 'data science' from 2013 to 2019. The turning point occurred in 2019 when the interest in 'data science' outperformed that in 'big data', indicating a focus shift towards the scientific investigation of data.

Practical applications of big data analytics can be found in many industries [53]. In finance, it can be used to detect frauds, evaluate investment risk, and facilitate transactions in stock markets. In healthcare, the digitalization of health records is leading to big data concerning patients, which can be used to evaluate treatment effectiveness, make predictive prognosis, personalize drug prescription, uncover a drug's side effects, etc. In manufacturing, big data analytics is useful for quality assurance, supply chain management, warehouse management, energy management, predictive maintenance, and so forth. Concerning public safety, government can employ big data analytics to monitor illegal activities on the Internet and discover potential risks to public security.

In the context of engineering design, big data analytics is most applicable to the customer domain and service domain, which coincides with the value creation pattern suggested by the Smiling Curve in management. On the one hand, a huge volume of data about customer demographics, preference, journey, and demand has been generated on the Internet and Mobile Internet through e-commerce platforms, search engines, user-generated content (e.g., online customer reviews and product review videos), etc. Such big data can enable designers to deepen understandings of target customers based on which to make more informed decisions concerning market segmentation, targeting, and positioning. On the other hand, big data analytics has been increasingly employed in smart factories for anomaly detection, sequential and non-sequential process modelling, pattern recognition, trend prediction, and so forth [54].

5   Integration of Cyber-Physical Data

5.1   Virtual Reality

Virtual reality (VR) can simulate a real-world experience through digital means, being situated in a virtual environment. VR is characterized by a unique way of cyber-physical integration (i.e., situating physical users in a virtual environment). In a typical setting of virtual reality, users wear a VR headset to interact with a virtual environment as well as various objects within the environment. A typical VR headset consists of key components such as sensing system (e.g., gyroscope and motion sensors) for detecting body motions, display system for presenting the virtual environment, tracking system (e.g., camera-based with constellation or scanning laser with photodiodes) for tracking body motions with different degrees of freedom, and computing system for processing data.

Practical applications of virtual reality can be found in various industries [55]. In entertainment, VR can be used to create an immersive user experience in video game, 3D movie, art exhibition, theme park, etc. In education, VR presents a cost-effective and visualized way of educating students to perform complex tasks in different environments [56]. In healthcare, VR can be used to enhance medical training, familiarize patients with an unfamiliar medical environment, situate physical therapy and rehabilitation in virtual environments, even treat certain diseases (e.g., post-traumatic disorders), etc. In manufacturing, VR can be used to facilitate worker training and conduct workplace analysis [57].

Virtual reality is especially applicable to the scenarios of data-driven design that involve a high degree of dynamic complexities within a changing environment. In particular, VR is proven effective in preparing human operators for an unknown or unfamiliar environment for the sake of safety, collaboration, and productivity. VR can be leveraged to support designers as well as customers to visualize, evaluate, and compare different design alternatives, being situated in a particular virtual environment. Furthermore, data collected from VR can be analysed by designers to understand design affordance (i.e., how a product may be used by different users in an open-ended environment within and beyond the original product-user interactions intended by the designers).

5.2 Digital Twin

Digital twin (DT) refers to the digital shadow/replica of a physical object (i.e., system, device, process) that is constructed in the digital world. Through continuous connection, communication, collaboration, and coevolution between the physical object and its digital twin, more informed decisions can be made [58]. From the technological perspective, DT is characterized by a holistic synthesis of various enabling technologies concerning different dimensions such as connection, service, and data [59]. In particular, the data dimension of DT involves enabling technologies for data acquisition, transmission, storage, computation, visualization, and fusion. Practical applications of DT can be found in industries such as agriculture, manufacturing, smart city, supply chain, healthcare, engineering management, energy, and so forth [60]. DT is proven useful for different stages of a product's whole lifecycle including design, production, distribution, and service.

The specific applicability of digital twin to engineering design is elaborated in Chap. 8 of this book. The reason why DT is reiterated here is because it serves as a key enabling technology for cyber-physical integration. DT can be regarded as a digital intermediator through which data generated by means of simulation (i.e., reflections of the digital world) and by means of IoT/IIoT (i.e., reflections of the physical world) can be seamlessly integrated. On the one hand, in comparison with the traditional simulation, DT can incorporate the real-world and even real-time data to enhance the accuracy and integrity of the simulation results. On the other hand, in contrast to data collected from IoT and IIoT, DT makes it possible to simulate certain scenarios that are too dangerous, complex, or unpredictable to instantiate and test in the physical world.

It should be made clear that, given the high cost and complexity of developing, operating, and maintaining a functional DT system, the paradigm of digital twin-driven design is by no means universally applicable to every scenario of engineering design. DT is particularly suitable for the data-driven design of high-value assets and complex systems for the industrial market such as aircraft engines, autonomous vehicle, smart building, smart grid, etc. For such asset/systems, it is oftentimes beneficial and necessary to develop a unique DT system to reflect the corresponding physical system according to a distinctive one-to-one mapping. On the other hand, DT can still be exploited to enhance the (re)design of medium-value and even low-value products for the consumer market, such as various smart appliances accommodated in a smart home. In that case, a generic DT system may be developed to monitor, simulate, and control the behaviours of numerous analogous products, which can be enabled by DT to learn from each other's behaviours.

5.3 Wearable Devices

Wearable devices refer to a particular kind of smart device that are worn on or near the skin of wearers. Typical examples of wearable device include smart glasses, smart shirt, smart bracelet, smart watch, smart shoes, smart ring, etc. Different from mobile devices such as smartphones and tablet computers, a wearer typically considers a wearable device as an integral part of his/her body. As a result, wearable devices make it possible to directly collect personal data about target customers. Data obtained from wearable devices can provide insights concerning customer preference, habit, health, etc. It is therefore imperative to develop extra mechanisms to protect the data privacy and security of wearable devices.

Wearable devices are generally classified into the category of consumer electronics. Wearable devices can lead to immense values in healthcare, where they can be used to collect a user's health data on body temperature, heart rate, blood pressure, footsteps, time on exercising, amount of calories burned, etc. Based on the analytics of such data, wearable devices can enable users to monitor their personal health, evaluate the potential risk for a certain disease, assess environmental impacts on personal health, and predict health changes over time. Furthermore, wearable devices can also be used in industrial scenarios such as smart helpmates for worker safety protection in factories, smart glasses for direct information tracking in transportation, and smart watches for health monitoring in mining.

6 Integration of Enabling Technologies

Data obtained from different enabling technologies can benefit engineering design and product development in different ways. Data obtained from virtual reality is most useful for reflecting the interaction between customers and environment, in particular, how and in what ways a customer would behave in an unfamiliar environment. Data obtained from the Internet and Mobile Internet can reveal customer demands, behaviours, preferences, journeys, and interactions in the virtual world. Data obtained from the Internet of Things (IoT) can indicate product behaviours, performance, and interactions in the physical world. Data obtained from wearable devices can reflect customer states, behaviours, and journeys in the physical world. Data obtained from digital twin can reveal cyber-physical interactions between product and environment.

For a data-driven design project, various enabling technologies can be integrated both vertically and horizontally. With respect to the vertical integration, for example, data obtained from the four kinds of Internet (i.e., the Internet, mobile Internet, Internet of Things, and Industrial Internet of Things) can be integrated to support interrelated design operations. Specifically, customer voices on the Internet can enable designers to deepen understandings of their target customers; the location data on the Mobile Internet enables customers to personalize the functions of a smart product against different contexts; and the temperature data collected by sensors can be communicated through IoT to adapt product behaviours. With respect to the horizontal integration, design data obtained from one source or channel can go through different data operations enabled by different technologies. For instance, the real-time data collected by sensors can be communicated through the IoT network, provisionally processed by edge computing, uploaded to the cloud, encrypted securely by blockchain, analysed by machine learning and deep learning, and compared against the simulation data based on digital twin.

An integrated solution of data-driven design can be modelled as a particular kind of System of Systems (SoS), which is composed of a selection of task-oriented systems [61]. Each of the before-mentioned enabling technologies can be regarded as a task-oriented sub-system. The design of such a system of systems by itself constitutes an intriguing while complicated design problem, for which many underlying design principles would apply. It should be noted that different enabling technologies are at different stages of their innovation lifecycles (i.e., innovation trigger, peak of inflated expectations, trough of disillusionment, slope of enlightenment, plateau of productivity). As such, different technologies by nature vary significantly with respect to their technological maturity, which should be taken into account of system integration.

7 Illustrative Examples of Practical Applications

This section presents a collection of practical applications of the before-mentioned enabling technologies for different facets of data-driven design. Each application is described with respect to the intended design problems, relevant data operations, and specific enabling technologies.

Recommender systems (i.e., an important branch of machine learning) are widely used in e-commerce to recommend suitable product or service to target customers based on the prediction of customer demand, preference, and satisfaction. In the context of product development, as shown in Fig. 1, recommender systems can also be adapted to recommend new functions to a target product [62, 63]. Functional formulation is one of the most challenging tasks in conceptual design. Figure 2 shows a systematic functional recommendation process through which the similarities between analogous products in tandem with online customer reviews are leveraged to recommend new functions to a target product. In practice, for example, relevant functions of a smart coffee machine may be recommended to a smart refrigerator. Online customer reviews, which are massively available on the Internet, constitute the data needed to power the functional recommender system. The key enabling

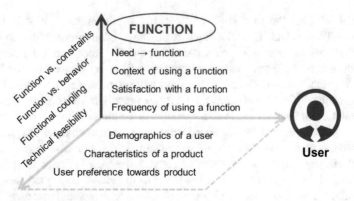

Fig. 1 Application of recommender systems in product development

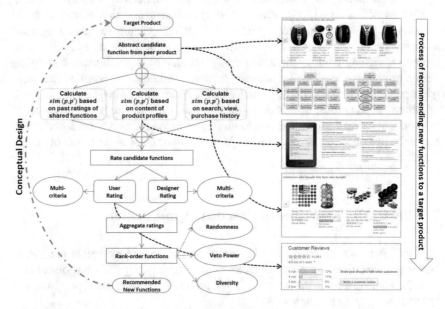

Fig. 2 Systematic process of functional recommendation [62]

technologies of a functional recommender system include machine learning, artificial intelligence, Internet, mobile Internet, and cloud computing.

In another application, blockchain can be used to enhance the trust among relevant stakeholders involved in a complex design customization process [30]. The lack of trust among stakeholders (i.e., between customer and manufacturer, between manufacturer and supplier, and between regulator and manufacturer) is a longstanding problem in product customization. As shown in Fig. 3, data generated in an iterative customization process can be recorded on both public and private blockchains. Various data blocks constitute not only a complete chain of decentralized consensus

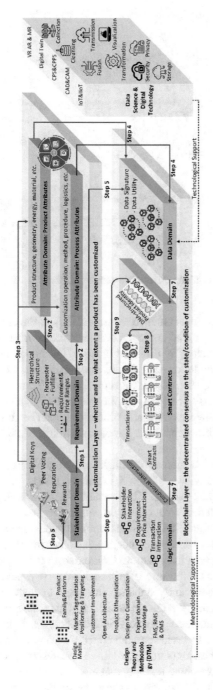

Fig. 3 Application of blockchain for smart customization [30]

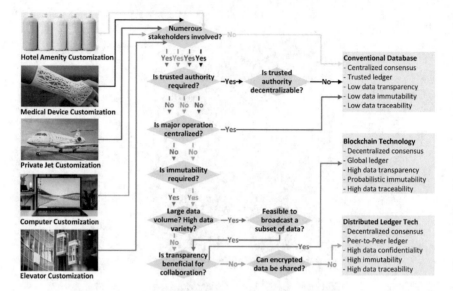

Fig. 4 Applicability of blockchain for different customization scenarios [30]

about the customization process but also a unique DNA of the customized product. Moreover, smart contracts can be deployed on the blockchain network to automatically enforce contracts concerning different facets of customization. Other relevant enabling technologies include IoT/IIoT for data collection and cloud computing for accommodating the blockchain network. It should be noted that, in consideration of the high cost, blockchain is not equally applicable to every scenario of customization. Therefore, Fig. 4 shows a systematic process that can be followed to evaluate the practical viability of blockchain for the customization of different products or services [30].

Sentiment analysis (i.e., a branch of artificial intelligence) can be employed to analyse online customer reviews. Figure 5 illustrates a structured framework that integrates sentiment analysis with the Kano model to automatically classify various product features into different categories (i.e., attractive feature, one-dimensional feature, and must-have feature) [64]. Besides, anomaly detection and novelty detection can be utilized to discover those abnormal customer reviews that may contain uncommon and contradictory information. Compared to the conventional practice of manually collecting and analysing data based on surveys and interviews, artificial intelligence can significantly enhance the scale of data analysis, the cost-effectiveness, and the efficiency of design decision-making. This specific application of data-driven design is enabled by artificial intelligence, machine learning, and the Internet (i.e., online customer reviews obtained from e-commerce platforms).

Data privacy is another major concern in data-driven design. Federated learning is an emerging paradigm of machine learning that enables numerous devices to collaboratively train a global model without exchanging or transferring raw data. As a

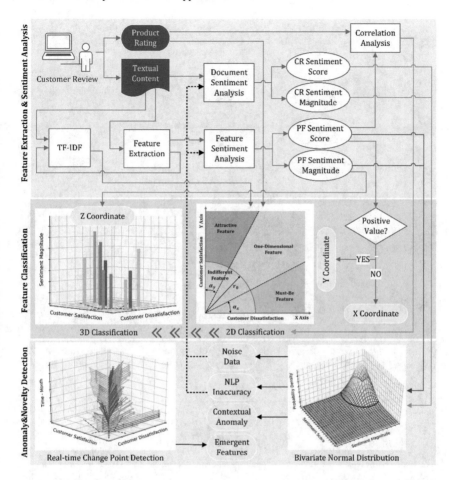

Fig. 5 Classification of product features based on sentiment analysis and anomaly detection

result, federated learning can be used to protect data privacy in product development [65]. Figure 6 illustrates a systematic process of smart product design through federated learning. The data can be collected by sensors and IoT from a smart product. The model training is twofold: edge computing is employed to train the model locally based on local data within smart products, whereas cloud computing is employed to optimize the model globally based on aggregated model updates. Therefore, this particular application of data-driven design is collectively enabled by machine learning, IoT, edge computing, and cloud computing.

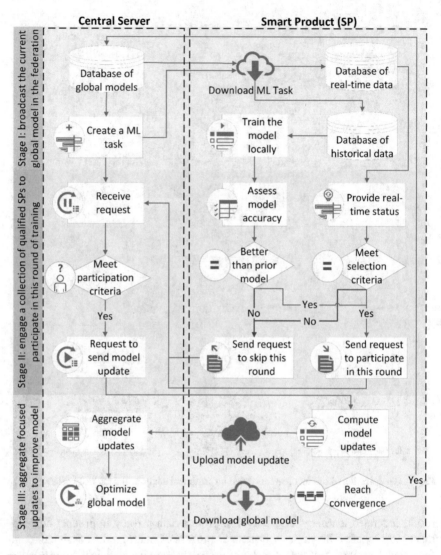

Fig. 6 Privacy-preserving smart product design through federated learning [65]

8   Conclusion

This chapter presents a collection of specific enabling technologies for data-driven design. These technologies are elaborated in correspondence to different data operations such as data acquisition, transmission, storage, computation, analysis, and cyber-physical integration. There exist inherent couplings among different enabling technologies. Therefore, the integration of various enabling technologies towards a functional system of systems by itself constitutes a complex design problem, for which a systematic design process should be followed to map the requirements of data operations to corresponding technologies.

References

1. Hirz, M., & Walzel, B. (2018). Sensor and object recognition technologies for self-driving cars. *Computer-Aided Design and Applications, 15*(4), 501–508.
2. Demiris, G., Hensel, B. K., Skubic, M., & Rantz, M. (2008). Senior residents' perceived need of and preferences for "smart home" sensor technologies. *International Journal of Technology Assessment in Health Care, 24*(1), 120.
3. Chae, M., & Kim, J. (2003). What's so different about the mobile Internet? *Communications of the ACM, 46*(12), 240–247.
4. Madakam, S., Lake, V., Lake, V., & Lake, V. (2015). Internet of Things (IoT): A literature review. *Journal of Computer and Communications, 3*(05), 164.
5. Asghari, P., Rahmani, A. M., & Javadi, H. H. S. (2019). Internet of Things applications: A systematic review. *Computer Networks, 148*, 241–261.
6. Gómez-Chabla, R., Real-Avilés, K., Morán, C., Grijalva, P., & Recalde, T. (2019, January). IoT applications in Agriculture: A systematic literature review. In *2nd International Conference on ICTs in Agronomy and Environment* (pp. 68–76). Cham: Springer.
7. Manavalan, E., & Jayakrishna, K. (2019). A review of Internet of Things (IoT) embedded sustainable supply chain for industry 4.0 requirements. *Computers & Industrial Engineering, 127*, 925–953.
8. Bhuiyan, M. N., Rahman, M. M., Billah, M. M., & Saha, D. (2021). Internet of Things (IoT): A review of its enabling technologies in healthcare applications, standards protocols, security and market opportunities. *IEEE Internet of Things Journal.*
9. Liao, Y., Loures, E. D. F. R., & Deschamps, F. (2018). Industrial Internet of Things: A systematic literature review and insights. *IEEE Internet of Things Journal, 5*(6), 4515–4525.
10. Chen, B., Wan, J., Shu, L., Li, P., Mukherjee, M., & Yin, B. (2017). Smart factory of industry 4.0: Key technologies, application case, and challenges. *IEEE Access, 6*, 6505–6519.
11. Lucke, D., Constantinescu, C., & Westkämper, E. (2008). Smart factory-a step towards the next generation of manufacturing. In *Manufacturing systems and technologies for the new frontier* (pp. 115–118). London: Springer.
12. Venters, W., & Whitley, E. A. (2012). A critical review of cloud computing: Researching desires and realities. *Journal of Information Technology, 27*(3), 179–197.
13. Durao, F., Carvalho, J. F. S., Fonseka, A., & Garcia, V. C. (2014). A systematic review on cloud computing. *The Journal of Supercomputing, 68*(3), 1321–1346.
14. Kavis, M. J. (2014). *Architecting the cloud: Design decisions for cloud computing service models (SaaS, PaaS, and IaaS).* Wiley.
15. Marston, S., Li, Z., Bandyopadhyay, S., Zhang, J., & Ghalsasi, A. (2011). Cloud computing—The business perspective. *Decision Support Systems, 51*(1), 176–189.
16. Xu, X. (2012). From cloud computing to cloud manufacturing. *Robotics and Computer-Integrated Manufacturing, 28*(1), 75–86.
17. Zhang, L., Luo, Y., Tao, F., Li, B. H., Ren, L., Zhang, X., ... & Liu, Y. (2014). Cloud manufacturing: A new manufacturing paradigm. *Enterprise Information Systems, 8*(2), 167–187.
18. Li, B. H., Zhang, L., Wang, S. L., Tao, F., Cao, J. W., Jiang, X. D., ... & Chai, X. D. (2010). Cloud manufacturing: A new service-oriented networked manufacturing model. *Computer Integrated Manufacturing Systems, 16*(1), 1–7.
19. Gao, R., Wang, L., Teti, R., Dornfeld, D., Kumara, S., Mori, M., & Helu, M. (2015). Cloud-enabled prognosis for manufacturing. *CIRP Annals, 64*(2), 749–772.

20. Abbas, N., Zhang, Y., Taherkordi, A., & Skeie, T. (2017). Mobile edge computing: A survey. *IEEE Internet of Things Journal, 5*(1), 450–465.
21. Shi, W., & Dustdar, S. (2016). The promise of edge computing. *Computer, 49*(5), 78–81.
22. Satyanarayanan, M. (2017). The emergence of edge computing. *Computer, 50*(1), 30–39.
23. Yu, W., Liang, F., He, X., Hatcher, W. G., Lu, C., Lin, J., & Yang, X. (2017). A survey on the edge computing for the Internet of Things. *IEEE Access, 6*, 6900–6919.
24. Zhang, J., Chen, B., Zhao, Y., Cheng, X., & Hu, F. (2018). Data security and privacy-preserving in edge computing paradigm: Survey and open issues. *IEEE Access, 6*, 18209–18237.
25. Chen, B., Wan, J., Celesti, A., Li, D., Abbas, H., & Zhang, Q. (2018). Edge computing in IoT-based manufacturing. *IEEE Communications Magazine, 56*(9), 103–109.
26. Taleb, T., Dutta, S., Ksentini, A., Iqbal, M., & Flinck, H. (2017). Mobile edge computing potential in making cities smarter. *IEEE Communications Magazine, 55*(3), 38–43.
27. Zhang, J., Deng, C., Zheng, P., Xu, X., & Ma, Z. (2021). Development of an edge computing-based cyber-physical machine tool. *Robotics and Computer-Integrated Manufacturing, 67*, 102042.
28. Xu, X., Weber, I., & Staples, M. (2019). *Architecture for blockchain applications* (pp. 1–307). Springer.
29. Zheng, Z., Xie, S., Dai, H. N., Chen, X., & Wang, H. (2018). Blockchain challenges and opportunities: A survey. *International Journal of Web and Grid Services, 14*(4), 352–375.
30. Liu, A., Zhang, D., Wang, X., & Xu, X. (2021). Blockchain-based customization towards decentralized consensus on product requirement, quality, and price. *Manufacturing Letters, 27*, 18–25.
31. Almada-Lobo, F. (2015). The Industry 4.0 revolution and the future of Manufacturing Execution Systems (MES). *Journal of Innovation Management, 3*(4), 16–21.
32. Cong, L. W., & He, Z. (2019). Blockchain disruption and smart contracts. *The Review of Financial Studies, 32*(5), 1754–1797.
33. Ali, O., Ally, M., & Dwivedi, Y. (2020). The state of play of blockchain technology in the financial services sector: A systematic literature review. *International Journal of Information Management, 54*, 102199.
34. Pournader, M., Shi, Y., Seuring, S., & Koh, S. L. (2020). Blockchain applications in supply chains, transport and logistics: A systematic review of the literature. *International Journal of Production Research, 58*(7), 2063–2081.
35. Alladi, T., Chamola, V., Parizi, R. M., & Choo, K. K. R. (2019). Blockchain applications for industry 4.0 and industrial IoT: A review. *IEEE Access, 7*, 176935–176951.
36. Drosatos, G., & Kaldoudi, E. (2019). Blockchain applications in the biomedical domain: A scoping review. *Computational and Structural Biotechnology Journal, 17*, 229–240.
37. Wang, Q., Zhu, X., Ni, Y., Gu, L., & Zhu, H. (2020). Blockchain for the IoT and industrial IoT: A review. *Internet of Things, 10*, 100081.
38. Lu, S. Y., ElMaraghy, W., Schuh, G., & Wilhelm, R. (2007). A scientific foundation of collaborative engineering. *CIRP Annals, 56*(2), 605–634.
39. Xu, R., & Wunsch, D. (2005). Survey of clustering algorithms. *IEEE Transactions on Neural Networks, 16*(3), 645–678.
40. Chandola, V., Banerjee, A., & Kumar, V. (2009). Anomaly detection: A survey. *ACM Computing Surveys (CSUR), 41*(3), 1–58.
41. Kaelbling, L. P., Littman, M. L., & Moore, A. W. (1996). Reinforcement learning: A survey. *Journal of artificial intelligence research, 4*, 237–285.
42. Kober, J., Bagnell, J. A., & Peters, J. (2013). Reinforcement learning in robotics: A survey. *The International Journal of Robotics Research, 32*(11), 1238–1274.
43. Haydari, A., & Yilmaz, Y. (2020). Deep reinforcement learning for intelligent transportation systems: A survey. *IEEE Transactions on Intelligent Transportation Systems.*
44. Oliff, H., Liu, Y., Kumar, M., Williams, M., & Ryan, M. (2020). Reinforcement learning for facilitating human-robot-interaction in manufacturing. *Journal of Manufacturing Systems, 56*, 326–340.

45. Wuest, T., Weimer, D., Irgens, C., & Thoben, K. D. (2016). Machine learning in manufacturing: Advantages, challenges, and applications. *Production & Manufacturing Research, 4*(1), 23–45.
46. Wang, J., Ma, Y., Zhang, L., Gao, R. X., & Wu, D. (2018). Deep learning for smart manufacturing: Methods and applications. *Journal of Manufacturing Systems, 48*, 144–156.
47. Miotto, R., Wang, F., Wang, S., Jiang, X., & Dudley, J. T. (2018). Deep learning for healthcare: Review, opportunities and challenges. *Briefings in bioinformatics, 19*(6), 1236–1246.
48. Zhao, Z. Q., Zheng, P., Xu, S. T., & Wu, X. (2019). Object detection with deep learning: A review. *IEEE Transactions on Neural Networks and Learning Systems, 30*(11), 3212–3232.
49. Simon, H. A. (2019). *The sciences of the artificial.* MIT Press.
50. Wang, L., Liu, Z., Liu, A., & Tao, F. (2021). Artificial intelligence in product lifecycle management. *The International Journal of Advanced Manufacturing Technology*, 1–26.
51. Arrieta, A. B., Díaz-Rodríguez, N., Del Ser, J., Bennetot, A., Tabik, S., Barbado, A., ... & Herrera, F. (2020). Explainable Artificial Intelligence (XAI): Concepts, taxonomies, opportunities and challenges toward responsible AI. *Information Fusion, 58*, 82–115.
52. Kim, S. G., Yoon, S. M., Yang, M., Choi, J., Akay, H., & Burnell, E. (2019). AI for design: Virtual design assistant. *CIRP Annals, 68*(1), 141–144.
53. Tsai, C. W., Lai, C. F., Chao, H. C., & Vasilakos, A. V. (2015). Big data analytics: A survey. *Journal of Big data, 2*(1), 1–32.
54. Gao, R. X., Wang, L., Helu, M., & Teti, R. (2020). Big data analytics for smart factories of the future. *CIRP Annals, 69*(2), 668–692.
55. Slater, M., & Sanchez-Vives, M. V. (2016). Enhancing our lives with immersive virtual reality. *Frontiers in Robotics and AI, 3*, 74.
56. Jensen, L., & Konradsen, F. (2018). A review of the use of virtual reality head-mounted displays in education and training. *Education and Information Technologies, 23*(4), 1515–1529.
57. Michalos, G., Karvouniari, A., Dimitropoulos, N., Togias, T., & Makris, S. (2018). Workplace analysis and design using virtual reality techniques. *CIRP Annals, 67*(1), 141–144.
58. Tao, F., Sui, F., Liu, A., Qi, Q., Zhang, M., Song, B., ... & Nee, A. Y. (2019). Digital twin-driven product design framework. *International Journal of Production Research, 57*(12), 3935–3953.
59. Qi, Q., Tao, F., Hu, T., Anwer, N., Liu, A., Wei, Y., ... & Nee, A. Y. C. (2019). Enabling technologies and tools for digital twin. *Journal of Manufacturing Systems.*
60. Tao, F., Zhang, H., Liu, A., & Nee, A. Y. (2018). Digital twin in industry: State-of-the-art. *IEEE Transactions on Industrial Informatics, 15*(4), 2405–2415.
61. Keating, C., Rogers, R., Unal, R., Dryer, D., Sousa-Poza, A., Safford, R., ... & Rabadi, G. (2003). System of systems engineering. *Engineering Management Journal, 15*(3), 36–45.
62. Liu, A., Lu, S., Zhang, Z., Li, T., & Xie, Y. (2017). Function recommender system for product planning and design. *CIRP Annals, 66*(1), 181–184.
63. Zhang, Z., Liu, L., Wei, W., Tao, F., Li, T., & Liu, A. (2017). A systematic function recommendation process for data-driven product and service design. *Journal of Mechanical Design, 139*(11).
64. Chen, D., Zhang, D., & Liu, A. (2019). Intelligent Kano classification of product features based on customer reviews. *CIRP Annals, 68*(1), 149–152.
65. Liu, A., Yu, Q., Xia, B., & Lu, Q. (2021). Privacy-preserving design of smart products through federated learning. *CIRP Annals.*

Printed in the United States
by Baker & Taylor Publisher Services